WHEN IT HURTS TO GROW

WHEN IT HURTS TO

JAN JOHNSON

VICTOR BOOKS®
A DIVISION OF SCRIPTURE PRESS PUBLICATIONS INC.
USA CANADA ENGLAND

Most Scripture quotations are from the *Holy Bible, New International Version,* © 1973, 1978, 1984, International Bible Society. Used by permission of Zondervan Bible Publishers. Other Scripture quotations are from (KJV) *King James Version* of the Bible.

Recommended Dewey Decimal Classification: 227.87
Suggested Subject Heading: BIBLE, N.T. EPISTLES: HEBREWS

Library of Congress Catalog Card Number: 90-23681
ISBN: 0-89693-197-8

1 2 3 4 5 6 7 8 9 10 Printing/Year 95 94 93 92 91

VICTOR BOOKS
A division of SP Publications, Inc.
Wheaton, Illinois 60187

CONTENTS

Introduction . . 7

1. Lean on Spiritual Giants . . 9

2. Make Heaven Your Home Now . . 19

3. Let God Be Your Parent . . 28

4. Let God Be Your Teacher . . 39

5. Seek Holiness, Not Happiness . . 49

6. Surrender to God's Discipline . . 58

7. Let Your Bitterness Melt . . 69

8. Watch His Discipline Help You . . 79

A Leader's Guide for Group Study . . 89

The more we grow in the Lord, the more we understand God and how wonderful life can be. Right?

To a certain extent, yes. But even though we love God, spiritual growth can be a struggle. We don't understand why God allows certain things to happen to us—our friends are diagnosed with cancer, our children fail in school, our marriages don't achieve what all the books say they should.

Sometimes, it seems as if God even initiates certain things in our lives. We fail in a project because of someone else's mistakes; a coworker doesn't like us no matter how hard we try; we never have enough money even when our income increases. How do we explain this?

Hebrews 12 is one of the few passages in the Bible that talks about the "discipline of the Lord." This passage tells us that God's discipline includes both the specific acts He initiates to teach us and the typical daily hardships by which we grow.

Hebrews 12 isn't a simple passage, but it's an intriguing one. At times, we will identify with down-to-earth phrases such as "losing heart"; other times we'll stretch to understand ideas such as "sharing His holiness."

In this book, we'll explore Hebrews 12:1-17 in depth. In each chapter, we'll study either a few verses or a theme from the passage. Sometimes, we'll single out an unusual phrase or explore examples of biblical characters who illustrate those verses. We may become friends with some Old Testament characters whom we never thought much about.

Examining the Background

This concept of "the discipline of the Lord" was probably not strange to the original readers of Hebrews. These Jewish Christians knew all too well how God had disciplined Israel throughout history. God taught, loved, and even literally fed His wayward child, Israel. The discipline of the Lord, though obscure to us, probably made perfect sense to them.

It's exciting to study this passage and think about having that kind of parent/child relationship with God too. We become His children in a deeper sense. We don't just ask for favors; we respond to His discipline. We surrender in completely new ways.

Using This Study Guide

This book may be used in several ways. Study it by yourself for a personal Bible study. If you study with a friend, it will deepen a friendship from simple surface social interaction to a true sharing of spiritual life.

This Bible study book is ideal for group study. You may want to form your own group or ask your pastor to help you start a group at your church or even suggest it to the leader of an existing Bible study.

Each chapter includes four sections:

Visiting the Doctor These Bible study questions allow us to consult with God, our "Doctor," for answers from His Word. Pray before you begin. Then read the suggested passage. If a question seems difficult to you, come back to it at the end of the study. If you study with a group, listen to how God is speaking to you through other group members. For some questions, of course, there is only one answer. Other questions ask for interpretation of Scripture and application of it. It is helpful to hear how the Word speaks to others. You'll probably identify with other group members and form new bases for friendships.

Binding Up Wounds This narrative section ties up loose ends from the Bible study. It offers real life illustrations of how the biblical principles work. It also includes specific suggestions for making the Scripture work in your life.

Following Doctor's Orders Do you remember this familiar adage: "Impression without expression causes depression"? This section offers suggestions for expressing the Scripture in your life.

It's tempting to consider this section as a "dessert" and skip it. Please don't. It's important to put the Scripture into action every time you read it. Leaving spiritual growth to chance means it probably won't happen.

Leader's Guide One page per study is included in the back of the book to help a group leader. First, the objective states in one sentence what the session is all about. If the group discussion becomes confusing, pull it back to this theme. Next, personal preparation helps a leader anticipate problems and know what to emphasize. Guidance is offered also on specific questions as needed. If you're using this book in a group Bible study, each group member should complete the above three sections for a particular session before meeting together.

CHAPTER ONE

VISITING THE DOCTOR

Is your spiritual life in a rut? Have you been overwhelmed by the hurts of the past and let your spiritual life stand still? Maybe you repeat motivating Scriptures over and over to yourself, hoping they'll jumpstart you.

The "Hall of Fame of Faith" in Hebrews 11 includes heroes who also experienced hurt, confusion, and anger. When we put ourselves in the places of these "pictures of righteousness," they inspire commitment in us. We pick ourselves up and ask God to work in our lives too.

Lord, thank You for supplying people in our lives that inspire us. We see how they look to You and we're amazed. We want to look to You the same way.

Look at Hebrews 12:1; Skim Hebrews 11.

1. Draw a star next to those in the following list of the "great cloud of witnesses" who suffered.

The universe
Abel, a righteous man who made honorable sacrifices
Enoch, a man who didn't experience death
Noah, who built an ark on faith
Abraham, the traveler to a foreign country
Abraham, the hopeful father
Abraham, who longed for a heavenly country
Isaac, who blessed his children for the future
Jacob, who blessed Joseph's sons as he died
Joseph, who foresaw the Israelites' exodus
Moses, who was saved in the bulrushes
Moses, who gave up life in Pharaoh's court in disgrace
The nation of Israel, who passed through the Red Sea and saw the walls of Jericho fall
Rahab, who welcomed righteous strangers
Others, who endured torture and death, sometimes escaping it

Place a checkmark by two or three of the above "witnesses" with whom you are familiar or perhaps feel a kinship. Pick one and write down why it's special to you.

Review Hebrews 11:13-16.

2. What do these witnesses have in common?

   How do their accomplishments compare with today's emphasis on excellence and success?

   Examine their successes listed above and choose one that most resembles any of your successes in the past.

   Which of the above "witnesses" do you need to dwell on most today?

3. In what ways can we adopt this "alien and stranger" attitude today?

4. Turn your thoughts to Christians who are alive today. Who is an example of faith to you in the way in which they have suffered and conquered? A friend, a pastor, a relative?

Read Hebrews 12:1.
"Let us throw off everything that hinders and the sin that so easily entangles."

5. Read the above phrase of Hebrews 12:1 in several versions, especial-

ly the *Revised Standard Version, Living Bible* and *Phillips Version.* Copy the phrase from the versions which you find most meaningful. Underline the most powerful word in each verse that you copied.

6. Give an example of how sin easily entangles us.

7. "Sin that entangles" may sound sinister while "everything" that "hinders" suggests that relatively innocent things can hinder us. List a few examples of the good habits, objects, or relationships that can hinder women in their growth.

   How do these "innocent things" hinder you—that is, how do they keep you from giving all of yourself to God? Have you ever been distracted from God's concerns because your thoughts kept wandering to these "innocent things?"

   What does the phrase "throw off" tell us about the energy required to rid ourselves of focusing on these "innocent things"?

"Let us run with perseverance."

8. What can Christians do when entangling sins and hindering habits dog them? Record the tool mentioned in each reference below and how it can help us:

   2 Timothy 3:16

   2 Corinthians 10:6

Matthew 6:34

1 John 1:9

Galatians 6:2

1 Thessalonians 5:17

9. Why do we sometimes reject these tools when we're experiencing trials?

Circle the tool that you now use the most. Put an X by the one(s) you need to use more often.

"The race marked out for us."

10. How do we know what the "race marked out for us" is? What is God's general will for all of us? (See 2 Peter 3:9, 1 Thessalonians 5:18, and Galatians 5:22-23 for specific ideas.)

Place a check mark by the above tools that help us determine God's will, i.e., the race marked out for us. Is it within God's will that Christians suffer? Find a biblical reference to prove your point.

BINDING UP WOUNDS

Does it ever seem as if everyone around you has an easy life? *Their* children behave. *They* take interesting vacations. *They* accept the mishaps in life.

We, on the other hand, are confused by our children. *We* scramble to take makeshift vacations. *We* shake our heads over the mishaps in life. *What were we supposed to learn from this?* we ask ourselves.

People tell us, "Don't take your problems so seriously." Yet we take them personally because we've come to know God as a Person. We know that He controls everything, so why doesn't He fix our situation?

After more advice and Scripture quoting, we swallow hard and try to forget it. Only we don't. Our prayer time becomes a loud sigh as we fall asleep. We no longer aspire to those life goals we set so long ago. There's a soft undercurrent of, "Why bother?"

We feel tired when we hear phrases such as "run with perseverance the race marked out for us." How do we keep up when it feels as if the wind is against us? The writer of Hebrews provides a list of people who persevered when they had obvious reasons to quit. The Bible reports that none of these witnesses received what had been promised (11:39). That makes them sound like losers, not winners. But we know they were spiritual winners—"They were all commended for their faith," and will one day "be made perfect" (11:39-40).

What were the survival skills of this faithful group? How did they trust in God when they had many reasons not to? It certainly wasn't their righteousness. When we look at the list, we can name mistakes that many of them made. Many of them committed sins that would disqualify them from serving on any church board today.

If we look at the comments the writer of Hebrews made about them (11:1-40), we find at least one attitude and one action that got them through: They hoped in unseen things and they obeyed no matter what. Both of these phrases sounds so spiritual. What do they mean?

When Believing Isn't Seeing

To "hope in unseen things" sounds like an impractical cliché. We've all known Christians who give vague "pie in the sky" answers to down-to-earth problems. How does hoping in unseen things differ from that? Specifically,

Moses was "looking ahead to his reward" (11:26).
Moses "saw Him who is invisible" (11:27).
Abraham was "looking forward to the city with foundations, whose architect and builder is God" (11:10).
They "were longing for a better country—a heavenly one" (11:16).
They "admitted that they were aliens and strangers on earth" (11:13).

The "witnesses" in Hebrews 11 understood God's power. They believed He would have the final say. They believed that joy and justice would finally prevail, that the future held rest and blessings. While I worry over the present drought, men like Ezekiel envisioned the eventual "showers of blessings" (Ezekiel 34:26). Their awareness of the unseen things gave them a hope that surpassed any earthly problems. They "did not receive the things promised; they only saw them and welcomed them from a distance" (11:13).

Hope, along with faith and love, greases our halting spiritual wheels. Brother Lawrence in his classic book, *The Practice of the Presence of God,* talks about their importance:

> "That all things are possible to him who believes; that they are **less difficult to him who hopes;** that they are more easy to him who loves, and still more easy to him who perseveres in the practice of these three virtues" (Brother Lawrence, *The Practice of the Presence of God,* Old Tappan, NJ: Spire Books, 1985, p. 25).

To some people, hope is a characteristic of wimpy people who aren't bold enough to do anything about life. "Gee, I hope so," sounds wishy-washy.

But hope is not the weak member of the faith, hope, and love trio; it's the outpouring of faith. Faith is believing that Jesus is coming again. Hope shows itself when we glance toward heaven in full anticipation of Jesus' return, thinking about what it will be like. While love means that I'm committed to someone, hope fleshes out that commitment so I stand by her and believe in her.

Hope means you don't plan to give up—ever. When I used to go through trials, I tapped my foot impatiently and said, "So, God, where are you now?" On an almost subconscious level, I thought that if life ever got too harsh, I would reject God. If He ever pushed me to the limit, I would give up on Him.

That's not what happened. One summer I laid in a hospital bed with five broken bones, two preschoolers needing care at home and my husband out of work. If I were ever going to give up, that would have been the moment.

Instead, I clung to God more than ever. He was my only hope. Giving up on Him would be a short-sighted thing to do. Who would throw away a canteen on a desert because she was mad at the desert? Who would discard an oxygen tank in the depths of the sea when it was her only tool for survival?

For the first time in my life, Job 13:15 made sense to me, "Though He slay me, yet will I hope in Him." That's when I knew that I would never give up on God. I needed Him too much. He had proven Himself to me so often by rescuing me from bad situations, from others' poor judgment, even from myself. He'd not given up on me even though I'd proved to be a slow learner. I had invested enough time in my relationship with God that I couldn't live without Him.

Some "logical" person could have listed for me all the reasons why I should give up on God, just as Job's wife did for Job (Job 2:9). Many days there was no money, no jobs available, and no way to get my kids to behave. But there was God and He was my unseen reason to hope. I believed that He loved me and would care for me. None of the calamities in front of me made any difference.

When Obeying Is Overwhelming

The other thing the brave cloud of witnesses did that kept them going was that they boldly obeyed, no matter what.

> Abraham "obeyed and went, even though he did not know where he was going" (11:8).
> Moses' parents "saw that he was not an ordinary child, and they were not afraid of the king's edict" (11:23).
> Moses "chose to be mistreated" (11:25).
> Others "conquered kingdoms" and "shut the mouths of lions" (11:33).
> Others "faced jeers and flogging," "were sawed in two; they were put to death by the sword" (11:36-37).

Their hope in unseen things persuaded them to do what was right regardless of the unknown factors. Noah built a boat in a world where it may have never rained. Abraham prepared to slaughter his promised son Isaac. Moses gave up a life of luxury in Pharaoh's court to defend a Hebrew slave. Through their hope, these "witnesses" saw the goal in the distance when others saw nothing.

We are bonded to these "witnesses" in a divine plan: "God had planned something better for us so that only together with us would they be made perfect" (11:40). They are examples of courageous obedience when others would have quit.

In our own churches today, we know Christians who have grown spiritually against all odds. They've survived cancer, divorce, and financial ruin. Despite their struggles, they have prayed, studied the Word, and worshiped God as they've walked through trials.

One such person for me was a choir director, Myrtle. I already admired her patient spirit and her sense of awe and worship. She talked about how God showed her that she shouldn't relax in her retirement, but fill the open position of choir director. In the choir, in the women's group, in the evangelistic Bible studies, she was ready to try new avenues to obey God.

When someone told me that one of Myrtle's adult daughters had disappeared and was later assumed dead, I was shocked. I studied Myrtle's face during choir practice. I listened to comments she made in meetings. There was no trace of bitterness.

I finally asked her about it. "We just don't know, but God knows." I couldn't believe her acceptance. Her life became a "picture of righteousness," a "proof text," so to speak, to show me that I could trust God even when I didn't know what was going on.

When Distractions Obscure the Picture

Somehow these "witnesses" let their unseen hope eclipse the sin and distractions around them. The writer of Hebrews used the word *entangles* to picture how our sin trips us up. For example, we may first nurse a grudge. Then our grudge involves us in gossip and exaggeration. Then we have to be careful not to gossip in front of certain people so we'll still seem spiritual. It gets more complicated as we go along.

"Everything that hinders" doesn't have to be deep, dark sin. It may simply be a habit, possession, or relationship that holds us back. We become distracted by too much time spent watching television, reading magazines, or working overtime. In themselves, they seem to do little harm, yet they switch our focus from our hope in unseen things.

Even intrinsically good things can hinder us. For example, Joan has two sons and the younger had polio as a child. As the boys were growing up, the older one grew distant from Christ because he felt such mercy for his younger brother with polio. In his quiet way, he'd become angry with God and it took him years to work this out. This boy had a merciful spirit, which is a positive spiritual quality. It was twisted by the enemy to be used against God.

Many times my heart for evangelism has distracted me. I get discouraged by Christians I consider lethargic. I've wondered why God doesn't light an evangelistic fire under a church.

Others of us question God because of an emotionally unstable home life or tragedies He allows us to suffer. We don't doubt His existence; we doubt His love. We may even have an invisible chip on our shoulder and decide that God will never help us.

Throughout this study, we'll look at past hurts and ask ourselves, *Why is my hope fragmented? Why am I holding on to this hurt? What lessons have I not learned?*

This first session has given us one key already. Dare to hope in the unseen ways of God. Like the "witnesses" in Hebrews 11, we may appear to be "losers" to others, but we have unseen advocates who help us and rearrange our situations for us. We aren't losers, we're just delayed winners.

FOLLOWING DOCTOR'S ORDERS

We all have events from our past that we don't quite understand. Sometimes they overwhelm us—a problem with a spouse, a parent, or a child. Your mother may have died recently or you've lost your job. Other events have faded but have not been forgotten—someone rejected us or we experienced a public failure. We may have dealt with these puzzles, but they still gnaw at us subtly.

We wrestle with these trials as Jacob wrestled with the angel (Genesis 32:22-32). We know we're supposed to grow from them and we've tried to do so, but still they linger.

Draw a symbol below of such a trial in your life. Use stick figures or simple drawings you might find on traffic signs.

Now draw a table underneath it to represent an altar. Pray and offer this trial to God. Ask Him to help you work through it during this study so that you will grow from it and reach new levels of maturity.

Think of someone in your church whom you know has withstood a trial such as surviving a spouse's death or overcoming alcoholism. Talk with them about it. Ask them questions such as:

- ☐ What kept you from giving up hope?
- ☐ What do you wish you'd done differently?
- ☐ How did that experience make you stronger?
- ☐ In what sense did you have to obey even when it was difficult?
- ☐ How did your confidence in the "unseen things" of God grow?

CHAPTER TWO

VISITING THE DOCTOR

Most of us are better at comforting someone else than comforting ourselves. When we're hurting, we try to remember those "solutions" we've offered others, but they slip away from us. So we grope for a life preserver to hang onto, a reward to get us through.

In Jesus' greatest hour of need, He focused on His future in heaven. Even as an earthly being, Jesus spent time praying and being alone with God. When He faced the Cross, He set the joy of heaven before Himself and endured the suffering.

Father, teach us to seek Your presence continually. Help us to look forward to Your coming and to life with You beyond the grave.

Read Hebrews 12:2.
"Let us fix our eyes on Jesus, the author and perfecter of our faith."

1. How can Christians "fix their eyes" on Jesus? How does it differ from talking or reading about Him?

2. The Greek word for "author" is *archegos,* meaning someone who takes a lead in anything. It's translated "prince" and "captain" elsewhere (W.E. Vines, *Vine's Expository Dictionary of New Testament Words,* Old Tappan, NJ: Fleming H. Revell, 1966, p. 88). What ways has Jesus taken the lead in your faith? How did He initiate it?

As the "perfecter" or "finisher" of your faith, how has Christ completed or matured your faith? How will He "end" your faith?

Read your answers to the last two questions. What reasons do they include about why you should fix your eyes on Jesus?

"Who for the joy that was set before Him endured the Cross, scorning its shame, and sat down at the right hand of the throne of God."

3. How does Jesus' method of enduring the Cross resemble the methods of the "witnesses" in Hebrews 11? (See 11:10, 16, 26-27.)

Read Philippians 2:8-11.

4. The "joy that was set before [Christ]" may have included His being seated at the right hand of the throne of God. What other events in this passage will occur that could have been a part of the joy?

Skim Matthew 26:57–27:56.

5. What events in this passage would have been causes for shame and heartbreak for Jesus?

   What evidence does Matthew 26:64 offer that Jesus was concentrating on the "joy set before Him" during His trial too?

Read Colossians 3:1-4.

6. What do these verses tell you about Christ's heavenly existence?

Read Isaiah 6:1; Matthew 19:28; 25:31; Revelation 22:1-5. Describe the throne of Christ (or if you like to draw, sketch the scene from these verses). What impresses you the most about it?

7. What does Christ look like in His glorified state? (For help, see Matthew 17:2; Mark 9:3; Luke 9:29.)

8. What will happen to those who accept Christ? (See also 1 John 3:2; 1 Corinthians 15:51-54.)

9. When is it most easy to set your heart and mind on the "things above"? Most difficult?

Read Acts 7:54-60.

10. What vision did Stephen have as the Sanhedrin turned against him?

    What was Stephen's condition at that moment? Was he full of fear?

    How did his listeners respond when he spoke of his vision? (v. 57)

11. How do verses 49-50 also show that Stephen had "heaven on his mind"?

Read Hebrews 12:3-4.

12. Paraphrase these verses in the simplest language possible.

13. Why do you think that focusing on Jesus' majesty helps us avoid growing weary and losing heart?

14. List below the reasons you give up on projects, people, and causes. To the right, write why Christ's example of endurance and His presence on the throne in heaven can bring hope to those situations. The first one is done for you.

REASONS	HOW CHRIST'S EXAMPLE GIVES HOPE
1. I feel inadequate.	God gives confidence and strength.
2.	
3.	
4.	
5.	

The repetition of the words *endured* and *resisted* suggests that it's dangerous to give up too soon. When have you done that?

15. Which of the "Hall of Fame of Faith" heroes in Hebrews 11 struggled against sin to the point of shedding blood?

How do you think Jesus was able to stay focused on heavenly realities and on the sacrifices of the prophets and others who went before Him? (See Mark 1:35-37; Matthew 4:2.)

BINDING UP WOUNDS

What does it mean to be focused on heaven?

The stereotype of heavenly-mindedness is someone who does nothing for the millions of starving and homeless people—who simply says, "God will take care of those people." These people are accused of being "so heavenly-minded they're no earthly good."

It doesn't have to be that way. Christ focused His mind on heaven, but He also took earthly concerns seriously. He cared for the hungry and the homeless and met their needs with a compassion that was so intense it had to be heavenly.

Focusing on heaven can motivate us to grow; it helps us endure. But how can we cram thoughts of heaven into our already busy days?

The Two-Track Mind

The shame of Christ hanging on the cross seems miles apart from the glory of His sitting at the right hand of the throne of God. Yet Jesus managed to envision the one while enduring the other. He put Himself in "two places at the same time," as we call it. We can't imagine two such opposite places.

The Scripture doesn't say that Jesus joyfully endured the Cross, but that He endured it because of the impending joy. He saw Himself as a citizen of another kingdom. He saw Himself united with God. We, who also have the hope of heaven, can also set this joy before ourselves.

It doesn't sound easy, does it? Yet many women are doing this to endure some of the most painful moments of their lives.

Perhaps you or a friend have taken childbirth classes in which you were taught relaxation techniques to ease the pain of childbirth. Teachers of these classes usually encourage couples to pretend that they're resting in a relaxing, peaceful spot they've enjoyed in the past.

For the deliveries of my two children, I recalled a shady patch of grass in beautiful Yosemite National Park in California. During previous summers we had camped there, so I knew it well. From that spot, I viewed majestic peaks as I turned the pages of whatever book I was reading. Within the space of twenty months, I spent lots of time rehearsing that scene for the two births.

Two years after the second birth, I was in a traffic accident, trapped in the car's wreckage. Without realizing it, I reached back into my past and lapsed into my "birth dreamland." There in the night with sirens shrieking and paramedics using crowbars to get me out, I laid my head against the headrest and imagined the trees in Yosemite. Even with six fractured bones, I don't remember feeling much pain. I'd spent all those months practicing my escape to my peaceful spot and when a painful situation occurred unexpectedly, my escape kicked in.

Now I'm practicing "heaven." I read Scripture about heaven and think of how it might look with Jesus on the right hand of God. I imagine Jesus as He was described at the Transfiguration. Before you dismiss this as a far out idea or a New Age visualization, remember that I'm not making anything up. You can put these scenes together with what you find in the Bible.

I'm practicing so that when the pain of rejection from family, friends, and those to whom I minister becomes too much, "heaven" will kick in. I can endure the pain in the perspective of heaven. When my physical discomforts make me feel like barking at the world, I can look forward to my future—heaven.

It's so different from my "birth dreamland." That spot in Yosemite isn't so pleasant on cloudy or snowy days. The maintenance people may have decided to set a huge trash dumpster there. I'm not even sure I could find it again now.

But heaven always exists. The throne of God is permanent. Someday the earth, including my haven in Yosemite, will be destroyed, but the throne of God goes on forever. Heaven is a future that is promised to me. I liken myself to my friend who gazes with pride at the land she and her husband bought for their retirement home. I'm viewing my "retirement" through the lens of Scripture.

Tuning into Heaven Now

Besides trying to picture heaven through the description in Scripture, we can adopt attitudes that will help us accept the reality of heaven.

Understand the impermanence of life on earth. The "cloud of witnesses" in Hebrews 11 "admitted that they were aliens and strangers on earth." Abraham never bought land in Canaan until he needed a burial site for Sarah (Genesis 23:1-20). God had promised him the land as part of the covenant, yet Abraham waited many years to invest his money in even a small part of it. Why? Perhaps because "he was looking forward to the city with foundations, whose architect and builder is God" (Hebrews 11:10).

Like Abraham, we can devote our time, money, and energy to spiritual things. We spend time steeping ourselves in Scripture. We spend money on missions, not mansions. We spend our energy practicing Christian virtues and building up other Christians. We know that these efforts will matter in the place we're heading for.

Think about how aliens and strangers behave. They depend on their own kind of sustenance; they practice their own customs; they stick together. In the same way, I need to work through my faith with my brothers and sisters in Christ. I need to confess my sins to them. I need to let them love me.

This means that our values and views and actions will differ from others around us. This doesn't surprise us. In fact, we use it to remind ourselves of our temporary visit here on earth.

View heaven as part of our future. When you think future, think heaven. As Americans turn 55, they start reading about retirement benefits. They think about relocating. They make decisions about possible dilemmas in the future. What if they become helpless? What if they're hooked up to a respirator?

My friend, Pearl, who's 78, talks about her future in heaven using down-to-earth terms. She anticipates being rid of back pain and seeing her deceased husband again. At half her age, I forget that heaven is my future too. I have no idea when I will die and enter heaven, but I'm beginning to consider my traveling plans. Why not have the fun of anticipating heaven the way I anticipate next year's trip to the mountains?

Talk about the reality of heaven. Heaven becomes more real to us when we talk about it in conversation. For example, when I came home from visiting a friend in the hospital, my children were full of questions. I explained that my friend (their friends' mother) had cancer.

"Will she die?" my children asked. I told them I didn't know.

"If she dies, will she go to heaven?" they asked.

"Yes, and it will be a wonderful place," I began. Then I talked about every passage I could think of that described heaven. We read in Revelation about the throne of God and the crystal river of the water of life. I explained how God would wipe away every tear from our eyes. There would be no more death or mourning or crying or pain (Revelation 22:1, 3; 21:4).

As I heard the words come out of my mouth, I was stunned. It startled me to hear myself say out loud that I believed in a place so unearthly.

After all, no one I know (except Christ) has ever seen heaven. It seems so other-dimensional, so abstract. I'd thought about heaven, but never talked about it as if it were real.

We need to hear the words of faith come out of our mouths. The more we speak forth about something, the more we believe it ourselves (2 Corinthians 4:13).

Value the unseen as we do the seen. The urgency of earthly tasks tugs at me. It's so difficult to have devotions when the end table next to me is cluttered or dusty. I forget that money and time given to God is as important and productive as money and time spent buying groceries. In the long run, these unseen priorities are more important. As strangers and aliens, we'll differ from the rest of our culture:

☐ We'll encourage our children to choose a career based on God's leading rather than the amount of money or prestige it brings.

☐ We'll know the joy of wearing an old sweater because we gave additional "sweater money" to our church's homeless ministry.

☐ We'll know that 20 minutes on the telephone with a hurting person counts far more than 20 minutes of tidying up our living room so that somebody will be impressed with our beautiful home.

Set aside moments with God. Jesus' life was characterized by sessions of solitude, fasting, and prayer (Mark 1:35-37; Matthew 4:2; 14:22-23; 26:34-44). By these means, He aligned His whole self, including His bodily desires, with God. All these hours training His mind, body, personality, and spirit to focus on God seems to have made it automatic with Him. His expert skill at "abiding" in God prepared Him to endure the Cross with the joy of heaven set before Him (John 15:1-4).

We are fooled by the apparent uselessness of abiding in God. What good does it do to "be still and know that I am God" (Psalm 46:10a), to fix our hearts on God (Psalm 112:7-8). Yet as we still the inner voices of our desires, we seek God and begin to replace those voices with His desires for us.

Just as Jesus built His inner life with God, so can we. Prayer, meditation, and Bible study aren't just for the super spiritual; they're for all of us who believe. They explain how Jesus could endure the Cross by concentrating on the joy that was set before Him.

FOLLOWING DOCTOR'S ORDERS

Give some concrete expression to your understanding of heaven by doing one of the following activities:

1. Pick a hymn or Scripture song that expresses hope in heaven and sing it every day ("I Will Sing the Wondrous Story").

2. Skim the Scriptures mentioned about the throne of God and Christ and draw a sketch to illustrate.

3. Journal about what you already know about heaven. Write your impressions of it:

   —What color does heaven make you think of?
   —What texture does heaven make you think of ?
   —What shape does heaven make you think of?

4. If you have children, explain to them what you know about heaven and why you think you'd like it there.

5. Choose a sound you hear everyday (telephone ringing, baby crying, the reminder tone of a computer or appliance). Let it remind you, whenever you hear it, that God is on His throne.

CHAPTER THREE

VISITING THE DOCTOR

"This hurts me more than it hurts you."

When we were young and our parents made this statement, it sounded ludicrous. *How could our parents know how real our pain was?* we wondered. *Why were they inflicting it on us?*

Now we use this stock phrase to discipline our own children when the turmoil inside our hearts overwhelms us. We want our children to stop their rebellious behaviors. We want them to grow into mature adults. Yet we feel bad about taking away their prize toys or spanking them. We don't want them to hate us.

As Christians, we ask God the same questions. Why are you allowing this? Don't you know that this hurts? We forget or don't understand that He is a concerned parent who is eager for us to mature.

Lord, prepare us today to see You as the merciful, holy parent that You really are. Help us set aside previous notions of Your discipline and study Your Word with new eyes.

Read Hebrews 12:5-6.

1. What is your first reaction to this passage?

    Fear of punishment?
    Dread of the unknown?
    Trust that God is working?
    Other:

2. What clues do verses 3 and 5 give us about the spiritual and emotional condition of the recipients of this book?

What situations cause you to lose heart?

Do you lose heart as indicated in verse 5 when you're rebuked (confronted) and feel convicted? Why or why not?

Recall a time recently when you felt as if you had "lost heart."

Read Revelation 3:14-22.

3. For what did Christ rebuke the church at Laodicea (through its angel)?

4. How did God feel toward Laodicea? (v. 19)

   What did He advise the church to do?

   What would be the results of surrendering to God?

5. Have you ever sensed a similar rebuke from God for yourself? If so, when?

Read Hebrews 12:7-10.

6. What is the role of hardship in our lives?

7. How does God prove Himself as a responsible, caring parent other than through His discipline? How does this make His discipline easier to accept?

   What does this passage say about people who are exempt from discipline?

   What does verse 9 tell us is the best response to God's discipline?

   What are some of the results of submitting to the Father of our spirits?

   Write below three statements that reflect a submissive spirit to God's discipline. (Draw from other verses if you wish.) For example, one might be, "I understand that this trial will make me a stronger person."

Read Jeremiah 31:18-21; Hosea 11:1-4.

8. What kind of experience was it for God to discipline Ephraim (the 10 northern tribes of Israel)? Note that in verses 18-19 Ephraim is speaking while God is speaking in verses 20-21.

   List here the words that show God's love for Ephraim.

9. What is the tone of God's speech in:

   Hosea 11:1-4

   11:5-7

   11:8-11

   How do explain the shift back and forth?

   If you're a parent, how does this resemble moments when you have to discipline your children, but it hurts you terribly? Describe your feelings in those moments.

Review Hebrews 12:5-10.

10. How does God treat His children?

    Is this similar or different from the way you treat or treated your children (or would treat them, if you had children)? In what ways?

11. How does God feel toward His children (v. 6; see also its Old Testament reference, Proverbs 3:12)?

12. Even if a parent disciplines a child in a kind, loving way, the child may still hold a grudge. Why?

    What set of feelings or beliefs about God do you have that affect your response to His discipline?

What attitudes do you usually see in parents who don't discipline their children? Are these attitudes compatible with God's character?

13. Do you think the discipline of the Lord is an angry action? Why or why not? Give evidence from this passage or other passages.

14. In general, do you find it easy to submit to discipline from earthly sources (such as earthly fathers, police, employers)? Why or why not?

    Do you respond to God the same way or differently?

15. The concept of discipline refers to more than just correction. It's a total training of the person. How is this concept reflected in the following verses?

    Proverbs 1:2-3

    Proverbs 23:23

    Titus 1:8

    In what ways was Jesus disciplined or taught in His life on earth? (Hebrews 5:7-9) How did He come to be obedient?

BINDING UP WOUNDS

When I gently remind my eight-year-old daughter, Janae, to clean her room, she sticks out her jaw as if I'd insulted her. Then she stares at me and says something like, "You don't love me."

Sometimes I play along with her. I say things like, "You're right. I must not love you because I never bake your favorite pie at Thanksgiving." (I do.) "I never take you to drill team practice." (I do.) "I never make dinner or teach you songs or have tickling fights with you." (I do them all!)

This rehearsal of the facts disarms Janae somewhat, but I still have to hug her and say that I love her. I believe it will take her years to understand that when I discipline her, it doesn't mean I'm against her.

I'm as suspicious of my Father God as Janae is of me. With each trial, I suspect that He's against me. I constantly have to recall from the Bible that God is a loving Being who cares deeply for me.

Not all Christians resent these trials as I do. Many are simply confused by them. They assume that since God is loving, He would never allow pain to come their way, not even if it will help them. The phrase, "the discipline of the Lord," disturbs them because they think that God wills our lives to be rosy. *How could an all loving Father be so hard on us?* they ask.

This session's Bible passage is one of the few that talks about the discipline of the Lord. It gives us insight into the side of God that may confuse or even scare us. Let's look at some of the characteristics of God's discipline.

The Discipline of the Lord . . .
Is Not Harsh or Unfair

The Anger Question The phrase, "the discipline of the Lord," evokes fear in many of us because we picture an angry God rebuking us. Let's see if this is accurate.

First of all, Hebrews 12:5-10 says nothing about anger. Discipline and anger don't necessarily go together. We link them because human nature is such that only the best of parents don't occasionally get angry as they discipline their children. God, however, is patient enough to discipline us without anger.

We can't project the unjust anger of humans on God. As we come to know God better by reading the Bible, we can set aside these images. Jesus, in particular, rebuked people with a strong gentleness.

☐ To the woman at the well, Jesus said, "Go call your husband." Only after she evaded His request did Jesus confront her with the truth: "You have five husbands and are living with one who isn't your husband" (John 4:16-18). Still, no anger is evident.

☐ Before Jesus told the rich young ruler to sell his riches to the poor, He looked at the man and loved him (Mark 10:21). Jesus' discipline was an extension of His love.

☐ Even when Jesus confronted the Pharisees with the seven "Woe to you" statements, He did it after they refused to listen to His repeated confrontations. It was a calculated warning, not a blast of ire.

Human parents discipline in anger because they have mixed motives. They want to correct their children, but they also want an outlet for their frustration.

Our God is the perfect Parent who doesn't get "carried away with justice" when He disciplines. He is the embodiment of all those qualities that make a good disciplinarian: patience, gentleness, and self-control.

The purpose of His discipline isn't vengeance, but righteousness and peace—our growth (Hebrews 12:11). His concern is not, "I'll give her what she deserves," but, "I'll give her what will help her grow."

God's Righteous Anger This isn't to say that God never gets angry or never disciplines in anger. He was often angry with Israel (Numbers 11:1). Yet God's anger is different from human anger. It is not abusive or unfair or as harsh as the anger of the mythological gods. Human anger is capricious—you never know what will set a person off. What is permitted one day may be a crime the next.

God is not that way; He is fair. His anger is an expression of His justice. "I will discipline you but only with justice," God told rebellious Israel (Jeremiah 30:11; 46:28).

God gets angry at sin because He's holy, not because He's unreasonable. His anger is like the righteous anger we feel about pornography or child abuse. (That's not to water down the intensity of God's anger. The writer of Hebrews goes on to say in 12:29: "Our God is a consuming fire.")

Few of us have witnessed truly righteous anger that was 100 percent righteous with no taint of self-seeking interest. It's so rare that it's difficult to even imagine. This is one of the ways of God we may not understand.

The words *rebuke* and *punish* in Hebrews 12:5-6 make God's discipline sound harsh. *Rebuke* means to confront, convict, or reprove. The writer of Hebrews could have used another Greek word *epitimao,* which means to give an undeserved rebuke if he intended that meaning. Perhaps you've never experienced this sort of firm, but fair, confrontation. This is what God is doing to us.

Punish in verse 6 refers to a symbolic scourging that disease and suffering brings (W.E. Vine, *Vine's Expository Dictionary of New Testament Words,* Old Tappan, NJ: Fleming H. Revell Co., 1966, p. 328). Once again, the writer of Hebrews didn't choose a Greek word for *punish* that involves vengeance. This word refers to a "chastening by the Lord administered in love to His spiritual sons." This fits with the idea of discipline being momentarily painful in verse 11.

Teaches Us God's Ways

The original Greek word for "discipline" used in Hebrews 12:1-17 refers to a child's training and instruction (W.E. Vine, *Vine's Expository Dictionary of New Testament Words,* Old Tappan, NJ: Fleming H. Revell Co., 1966, p. 183). His discipline is a refinement process. Wise parents today know that discipline is more than punishment. It includes teaching children right from wrong and guiding them to do what's right.

Spiritual discipline is necessary because we don't become instantly spiritual when we accept Christ. Instead, we have to grow. And one of the ways we grow best is through God's discipline.

It's as if God has two types of "lesson plans" in His discipline curriculum.

☐ Sometimes He initiates a specific action to correct His followers. For example, God disciplined Israel by forcing them to wander 40 years in the wilderness (Numbers 14:1-45); Christ disciplined Laodicea with a specific warning to repent (Revelation 3:14-19). In the same way, God may motivate a specific person to confront us about a wrong behavior.

☐ "Endure hardship as discipline" (Hebrews 12:7). God allows hardship, trials and tribulations in our lives to discipline us, in a broader sense. We know that He will prevent any trial that is too much for us, just as He limited Satan's attacks on Job (Job 1:12). He promises us that our trials will work for our good if we love Him and work for His purposes (Romans 8:28).

God may not originate this second category of discipline. These trials may occur because of our foolishness or someone else's. Our present unredeemed world doesn't work harmoniously. God probably doesn't initiate your mother's death or a rebellious teenage son's attitude or a church split the way He forced Israel to wander. But the results are the same: God uses trials to help us surrender to Him.

God uses His discipline to communicate with us. We learn His ways. We surrender to His will. We depend on Him more. That's why we often hear Christians say that they were closer to God when they were going through a trial than after it was over.

That dependence on the Lord results in a deeper bonding. We build a history together. We understand God's "track record" of helping us survive the pain. In this way, our trials are "classes."

A Sign of Our Specialness

God's discipline is a sign that He considers us "true sons" or children. He's grooming us for our inheritance in heaven.

Think of the special relationship that exists between children and the people who discipline them. When I've disciplined my friends' children, they accepted it in most cases. They know that their mom and I are close friends. I've spent time talking to these children and they've watched me prove myself as a fair disciplinarian with my own children.

In one case, I overstepped my bounds. This boy hadn't seen me interact much with his mom or with my own children. I hadn't spoken with him often. When I corrected him, he looked at me as if to say, *You don't have any right to discipline me.* In a way, he was right. We had no history together.

God has earned the right for us to respect His discipline through His other excellent parenting skills. He provides for us, He offers us eternal life, He comforts us. This God who disciplines us is the same One who has rescued us out of bad relationships and dangerous situations. We can trust Him.

Even Jesus went through refinement. "During the days of Jesus' life on earth, He offered up prayers and petitions with loud cries and tears to the One who could save Him from death, and He was heard because of His reverent submission. Although He was a Son, He **learned obedience** from what He suffered, and, once made perfect, He became the source of eternal salvation" (Hebrews 5:7-9). If Jesus, who committed no sin, had to learn obedience, how much more we must need it.

Saves Us from Death

Another parenting cliché: "You don't understand this now, but you will someday." Teens don't like curfews; babies are intrigued by electric sockets. Parents are people who understand dangers and intervene.

God understands the danger of final judgment and disciplines us to help us escape it: "When we are judged by the Lord, we are being disciplined so that we will not be condemned with the world" (1 Corinthians 11:32).

Submitting to this discipline not only helps us escape eternal death but it brings spiritual life. "He who heeds discipline shows **the way to life,** but whoever ignores correction leads others astray" (Proverbs 10:17). How much more should we submit to the Father of our spirits and live!

This added dimension of spiritual life is evident in those whom God disciplined. Abraham, who was willing to offer His son to God, was called "God's friend" (James 2:23); Moses was said to be "more humble than anyone else on the face of the earth" (Numbers 12:3). The discipline of the Lord deepened their spiritual lives.

The discipline of the Lord is not a policeman's club. God is not hiding in the bushes waiting for us to goof so He can pounce on us. Neither is God a Santa Claus type either. He doesn't grant every whim. God is one who walks along side coaching us (John 16:8-11). He uses His discipline to guide us into paths that will bring us closer to Him.

FOLLOWING DOCTOR'S ORDERS

Think for a few minutes about how God's discipline differs from the way your parents disciplined you as a child. Then draw a vertical line down the center of the page. Write some of the characteristics of God's discipline on the left. Then on the right, write Yes, Sometimes, or No to indicate whether your parents disciplined you the same way most of the time, some of the time, or little of the time.

Here are some characteristics of God's discipline as starters: fair, forgives easily, instructive, desires repentance, confrontive, controlled, helps us avoid judgment.

Reread what you've written and write a short prayer to God emphasizing your appreciation of His discipline.

CHAPTER FOUR

Let God Be Your Teacher

VISITING THE DOCTOR

Have you ever wondered why bad things happen to you? Do you sometimes even wonder if God is picking on you?

No, you say, you know better than that. Yet you wonder why these trials come. Is God showing you a better path or slapping your wrist?

When circumstances sour and people fail us, we wonder what's going on. Why weren't our prayers answered? Why do certain people distance themselves from us?

We view these trials better if we have an adequate understanding of the "discipline of the Lord." Discipline makes most people think of spankings and withholding privileges. God's discipline also involves education.

This means that trials serve as the spiritual classrooms in the laboratory of life. They are tough internships in which we grow up. These hard knocks take the words on the printed pages of our Bibles and etch them in our hearts.

As we read about the "Lord's discipline," think about ways that God has trained or educated you.

Master Teacher, who loves me: Help me to receive Your teaching today as one who sits at the feet learning from her Father. Help me to listen and absorb the love with which You said these words.

Read Hebrews 12:5-6.

1. In what two ways does verse 5 imply that many people respond to the Lord's discipline?

Do you respond in either of these ways? How do you do that?

2. What qualities or activities are necessary to get the most out of a trial? See also 2 Timothy 2:15; 1 Thessalonians 5:17; Ephesians 5:20.

   Which of these is most lacking in your life?

Read Proverbs 3:11-12.

3. Some have subtitled Proverbs 3 "Further Benefits of Wisdom." Skim verses 1-10 to find out why.

   The writer of Proverbs commanded his reader not to do certain unwise things. What were they?

Read Hebrews 12:7-10.

4. How do you feel about teachers who don't discipline their students?

5. What does this tell you about the kind of teacher that God is?

6. How does education lose its effectiveness when people refuse to endure—or stick with it?

7. How does God "tutor" us (give extra help) as He disciplines us? (John 14:15-18; Romans 8:26)

8. Think about one or two of your favorite teachers in school. Were they lenient? How did they help you learn? When did you like them?

   What qualities did they have that are similar to God's ways?

Read Numbers 13:21-32; 14:1-4.

9. What report did most of the spies bring back about the Promised Land—the land God promised to Israel?

   What was their recommendation? (v. 31)

   How did the people respond? (14:4)

10. How had God prepared the Israelites to make the choice to enter the land? (Exodus 14:21; 20:21-22; Numbers 20:10-11)

Read Numbers 14:26-30.

11. How did God respond to the Israelites' fear?

12. How would this lapse of time (40 years—a generation) help the nation of Israel? How would it have better equipped them to be ready to conquer cities? To follow a leader into battle?

13. What do you think—since you can look with hindsight—of God's discipline of the Israelites? In what ways could it have been more severe?

14. What tasks do you shrink from out of fear or lack of trust in God's ability to help you?

    How do these become opportunities that slip through your fingers?

15. God educated the Israelites through this setback, this delay, this change of leadership. Has God disciplined you this way? How has it "educated" you?

BINDING UP WOUNDS

My wise friend, Katharine, once told me: "When you look back on your life, you will find that you learned more from the tough times than from the easy ones."

We both knew Katharine was talking about a former employer who continually criticized me. It especially hurt because he was an older Christian and I looked to him for leadership. I absorbed his criticisms and soon viewed myself as a failure.

When he hired another person and criticized her the same way, I saw that he was simply a critical person. I tried to mend my relationship with him, figuring that two Christians could work things out. It only made matters worse. We never resolved the conflict, so I found another job.

I resented this experience because I couldn't see any good in it. A few years later, I realized that I'd learned a few lessons. I learned and accepted that all Christians aren't perfect. I was more eager to please God instead of people. My sense of humor had helped me get along with disagreeable people.

Yet, even with what I learned, I bitterly resented having to suffer the consequences of my former employer's critical personality. I developed a spiritual "chip on my shoulder" because I saw this trial as an undeserved punishment.

Studying Hebrews 12 helped me understand that my experiences weren't punishments. More likely, they were part of the "discipline of the Lord."

This term involves more than just correction. The original Greek word for *discipline* means education. The word *discipline* is even paired up with *wisdom* at times (Proverbs 1:2, 7). Substituting *classes* and *education* for *discipline*, Hebrews 12:5-6a reads: "My son [child], do not make light of the Lord's *classes,* and do not lose heart when He rebukes you, because the Lord *educates* those He loves."

God disciplined Israel when He sent her into the wilderness for 40 years. There, Israel learned to eat manna from God and feed on the Lord's word. The Israelites saw how their clothing never wore out (Deuteronomy 8:1-6). As Israel trusted God to survive in the wilderness, she became ready to trust Him enough to conquer the Promised Land. In the same way, we first trust God in smaller situations, then in larger ones.

When I saw my trials as education, I understood that God wasn't punishing me. He was trying to impart wisdom about life to me. Even though I resented suffering for my supervisor's faults, I was willing to suffer to grow wiser about the Christian life.

I began picturing God as the great lesson writer in the sky, saying, "How can I help my children grow?" I think that my grumpy supervisor was part of the lesson plan. The hurtful situation showed me that criticism defeats while affirmation helps. Thankfully, my wise friend Katharine seemed to be the second part. An affirming person, she made me feel worthwhile once again.

And she was right about learning the hard way. As an advanced level critic myself, I recalled the image of my former supervisor when I began attacking my husband, my children, and my friends.

I'm doing it again, I'd tell myself. Then I'd contrast the supervisor with Katharine and choose to be as affirming as Katharine was. The hurt I'd experienced seemed to burn the lesson into my soul.

Here are some ways to become a better student of God's discipline.

Understand the Teacher's Methods

God is our teacher in the classroom of life, and He's a good one. He's not the kind who writes impossible assignments on the chalkboard and walks away. It's true that His disciplinary situations may ask us to do what seems impossible, but God goes the extra mile as needed. He not only answers questions in class but He also stays after school to help. He sends extra help to tackle each discipline experience. (This is the role of the Holy Spirit, to come along side us and counsel us—John 14:16.) He doesn't allow a trial unless He equips us to handle to it (1 Corinthians 10:13).

God demonstrated His teaching methods many times in the Bible. Has He disciplined you in ways similar to the following incidents?

☐ Jesus allowed Peter to deny Him, but He went the extra mile in receiving Him back (John 21:15-17).

☐ God allowed the Jerusalem church to be persecuted so that the Christians would scatter and spread the Word (Acts 8:1-4).

☐ God allowed Paul to keep his thorn in the flesh so he would feel the need for God's grace (2 Corinthians 12:8-9).

As a teacher, I know better than to make learning routine for my students. Then they don't remember what they've learned. If I make it interesting, they never forget.

Several times we've had intense moments where the class strained and struggled to guess a correct answer. I knew the discussion was taking up too much class time but I allowed it anyway. Once they discovered the answer, I knew they would never forget it.

God's discipline makes learning interesting. Our whole selves become involved as we struggle through His discipline with our hearts, minds, souls, and spirits.

Pay Attention in Class

We cheat ourselves when we "make light of the Lord's discipline" (Hebrews 12:5). Making light of it means we get comfortable being uneducated. We don't seek after what God is teaching us. We settle for "fast food" nourishment, instead of feasting on the Word and prayer.

As we grow, we learn to perk up our ears earlier to what might be in the heavenly air waves. We think through God's ways and even journal about them as David did in the Psalms. This helps us cut through surface attitudes and develop new layers of righteous longings. We become pliable so that God can work with us.

Sometimes we don't understand the importance of what we've learned until we encourage a fellow struggler with our "war stories." Our listeners marvel at how much we've learned. We marvel too because it's in reflecting on God's discipline that we finally understand it! Then the people who have criticized us, the friends who rejected us, or the tasks at which we failed don't seem so bad after all. We don't have to search every problem in panic for a moral. But as we reflect on events in our lives, we rest in God's comfort and wait for insight.

Watch Out for Learning Disabilities

We can't pay attention to God's lesson plan when we're busy focusing on negative attitudes. Let's call these attitudes "spiritual learning disabilities." For example, idolatry was a spiritual learning disability for the Israelites. Even after they were disciplined for worshiping the golden calf, the Israelites worshiped false gods in the Promised Land. Their attachment to idols not only turned them away from God, it ruined their nation. Altars to false gods built by Jeroboam cemented the division between north and south (1 Kings 12:25-33). Idolatry brought captivity to both nations (Ezekiel 5:8-12; Hosea 8:1-10). If Israel could have dealt with this spiritual learning disability early on, she might have saved herself some grief.

Here are a few common spiritual learning disabilities:

☐ *Covetousness* We have idols that distract us just as the Israelites did. How could it be God's will that we not have a new car now? How could He want us to live in a less prosperous neighborhood? How could He want our children to feel lonely?

These questions keep us from paying attention. Instead we surrender to the idols of wanting what's bigger and better, of appearing to be perfect women. These idols trip up our spiritual training. Our complaints and lack of gratitude blind us to what God is doing in our lives.

☐ *Criticism* The Israelites let their complaints block their vision of God's powerfulness. They harped on Moses; they blamed their worship of the golden calf on Aaron.

It's easier to assess blame than to endure a trial. We nag at the person in charge, we find fault with the participants, we reproach ourselves for every little mistake.

A critical spirit cuts our learning short as we focus on errors, especially those of others. Paul told the Romans, "For at whatever point you judge the other, you are condemning yourself, because you who pass judgment do the same things" (Romans 2:1). We waste our energy resenting others when we could be learning from their mistakes and looking for ways to bless them.

☐ *Self-pity* Just after their Sinai experiences, the Israelites complained that they were better off in Egypt where they had been overworked and beaten than traveling in freedom to the Promised Land (Numbers 11:1-6). Self-pity stunted their growth and prevented them from rejoicing in God's care for them.

Watchful parents spot their children's learning disabilities and start treatment for them as soon as possible. The treatment for these three spiritual learning disabilities is a grateful spirit.

Send Your Teacher Apples

Teachers in the frontier days made barely any money and the town's people often subsidized them with food. An apple on the teacher's desk was a juicy acknowledgment of what she meant to her students. It was good for the students too. It meant that they valued their teacher. We, by the same token, need to present God with apples of thankfulness. It curbs our learning disabilities and helps us see the Teacher for who He really is.

I haven't been quick to learn this. When my husband, Greg told me that his company looked as if it might shut down, I panicked I talked with my pastor friend John and he tried to walk me through potential problems. I fought him on each point, telling him why none of his solutions would work. Finally, I spit out a deep resentment: my husband had been out of work once before for 18 long months. It had been a time of both growth and misery. I still felt sorry for my husband and for myself that we'd had to go through it.

As I talked with John, I was so ashamed of my feelings of self-pity that I tried to sound a little more cheery. So I added a positive note at the end. "He went without work for 18 months, but we never missed a payment. Our house payments even went up, but we never missed a payment."

Then I became so quiet that John finally said, "Are you OK?"

My last five words, "we never missed a payment," echoed in my ears. I thought about how John had held a seminar on the Psalms in which the teacher pointed out how often the Israelites talked about the Exodus. Over and over, they spoke of how God had miraculously delivered them—especially in their crossing the Red Sea.

I'd had a Red Sea experience in never missing a house payment, but I had never viewed it that way. I'd spent my energies questioning why God would let my husband be without a job for 18 months instead of thanking Him for providing us with the resources we needed to survive. I'd never seen my "Red Sea experience" for what it was.

The realization that God had taken me through my own Red Sea empowered me. The next day a friend asked about Greg's job situation. When I told her it was uncertain, she said, "How are you taking this?"

I felt empowered once again and said, "God's gotten us through worse circumstances. I feel confident that He'll be there again."

I must have sounded awfully lofty because she giggled and said, "Are you faking this spiritual stuff?"

I laughed and said, "No. For today, for this moment, it's real. I believe God will stand by us."

And He did. I'm not so easily scared by broad Red Seas anymore.

FOLLOWING DOCTOR'S ORDERS

Think of the last trial you experienced, even something as small as misplaced insurance forms or a worn out refrigerator. Ask yourself these questions:

- ☐ What was God teaching me?
- ☐ How could I benefit from this trial?
- ☐ What's the worst that could happen to me because of this? How could that be beneficial?
- ☐ How did God protect me from much worse tragedy?

Next time a trial occurs, go through these steps. Then confess your panic to God and ask Him to take over. Do this as many times as necessary. Don't be impatient with yourself if it takes several times. This is normal.

Think about your spiritual learning disabilities. Are you prone to criticism or self-pity? How do these things keep you from growing?

Describe a recent trial you experienced. What did you learn from it? How would you be shortchanged if God had intervened and not allowed it to happen to you.

CHAPTER FIVE

Seek Holiness, Not Happiness

VISITING THE DOCTOR

If we're honest, most of us would like to lead a pain-free, happy life. We'd like for our circumstances to be cheery and the events of our life to flow. We'd choose a "happy face sticker" as our logo.

That's where the discipline of the Lord becomes a problem. Whether God singles us out for specific training or we have to work through our daily hardships, discipline can be painful. It's difficult to set aside our comfort and trust God to work for our good and bring us through.

Happiness, or more properly joy, comes as a by-product of knowing the Lord and His holiness. Growth is the long, but permanent, route to joy.

Teach us, Lord, to seek You first. Help us to set aside the temporal trappings of life that distract us from Your goals for us.

Skim Hebrews 12:1-13.

1. Make a list of some of the goals mentioned here for our spiritual lives.

2. Why do you think our spiritual growth is so important to God? Draw from Scripture you already know. Then consider 1 Timothy 2:4, 2 Timothy 1:9, and Romans 8:29-30 for some ideas.

3. Underline in your Bible or write on this page all of the "bad feeling" words you find in Hebrews 12:1-13. For example, "enduring the Cross" and "scorning its shame" don't create happy pictures.

The NIV translation uses the word *painful* in verse 11. What word(s) does your version use?

4. Verses 10-11 say that God's willful infliction of pain on us is good. Why?

5. In what ways does this verse imply that God's parenting is better than human parenting?

6. Some Christians resent God's discipline because they have mistaken ideas about God's protection and blessing. Read the following verses and see if you can explain why bad things happen to good people. Relate them as much as you can to the discipline of the Lord.

   Psalm 91:12

   Psalm 84:11

   Matthew 7:7

7. How can a Christian going through the discipline of the Lord relate to the following passages about joy?

   Philippians 4:4-8

   John 16:22

Read Habakkuk 1:1–2:1.

8. What was Habakkuk's first complaint about the wickedness of his own nation, Judah? (1:1-4) (Habakkuk lived in Judah's wicked last days just before she was about to be captured by the evil Babylonian empire.)

What was God's answer? (1:5-11)

9. What was Habakkuk's second complaint about the Babylonians as their conquerors? (1:12-17) Why didn't Habakkuk like God's solution to Judah's wickedness?

What was Habakkuk's tone? (2:1)

Read Habakkuk 2:2-20.

10. Write down a few phrases that describe why God would eventually condemn the Babylonians too. Many of them begin with "Woe to him."

Read Habakkuk 3:1-2, 16-19.

11. What does Habakkuk conclude about God?

12. When you consider that Habakkuk's society depended on fig trees, grapes, olive crops, sheep, and cattle for their livelihood, what is verse 17 saying?

Paraphrase verses 17-19 with contemporary culture and your needs in mind.

13. Look at the steps Habakkuk took in coming around to God's will. Have you had a similar experience?

☐ He complained because Judah was wicked and needed the Lord's discipline.
☐ He didn't think the instrument of discipline God chose to use was proper.
☐ He accepted God's judgment without bitterness.

BINDING UP WOUNDS

In my teen years, this phrase was popular: "Jesus is the answer." I believed this phrase meant that Jesus would solve every problem I ever had. Over the years, I've learned that knowing Jesus doesn't solve every problem, but that He equips me to deal with every problem. But, still I keep wishing He would solve everything for me. I can't seem to let go of the American dream of happiness. After all, it's part of the U.S. Constitution—we have the right to life, liberty, and the pursuit of happiness.

Also in my teen years, I heard a lot of Scripture about finding joy and fulfillment in Christianity. Either I was given joy-slanted messages or my joy hungry ears heard what they wanted to hear.

Happiness, or joy, is sometimes a by-product of knowing the Lord and His holiness. But a by-product isn't something you seek. If you're as preoccupied with being happy as I have been, you may have played games with God. For example, I knew that Matthew 6:33: "Seek ye first the kingdom of God and His righteousness; and all these things shall be added unto you." I assumed that "all these things" meant whatever I thought would make me happy. (Verses 30-32 indicate that "all these things" probably refer to food and clothing.) So I made a habit of seeking God first and then saying, "I've obeyed, so when do I get joy, God?" Without realizing it, I was still seeking joy, instead of seeking God.

My greatest struggle with accepting the discipline of the Lord has been to give up the quest for a cheery, well-ordered life. I like warm fuzzy get-togethers. I like well-planned days. I like it when clerks offer insightful help. I like to get my money's worth for what I buy. When these things happen, I get a surge of energy.

Too often, the discipline of the Lord brings struggle and sadness with it. It separates me from my comforts and blessings. His discipline then, contradicts what, in reality, is my life's goal: to be happy.

Over the years, God has dealt with me to change my perspective. If you struggle with this also, here are three truths to focus on.

God's Goal: Holiness, Not Happiness
The verse most often quoted to prove that Christians have a right to unending happiness is John 10:10: "I am come that they might have life, and that they might have it more abundantly" (KJV).

Jesus gives not only life, but life "to the full" (NIV). This harkens back to John's earlier words: "From the fullness of His grace we have all received one blessing after another" (John 1:16). The Christian life is full of blessings never known to non-Christians. The problem occurs when we assume that Christianity brings blessings only. You may think that this attitude isn't prevalent in your Christian circles, but think about comments you've heard similar to these:

"If I'm happy, I must be in God's will."

"If I don't sing with joy in a worship service (or at least smile some), I'm not a spiritual person."

Even though the Christian life is full of blessings, happiness cannot be our goal. A cheerful spirit cannot be the litmus test for whether or not we have faith. Christianity is not the fast track to happiness or a pop psychology or therapy to cope with life. It's a relationship with a Deity. Like all relationships, it has varied dynamics.

A quest for happiness, or even a minor preoccupation with it, distracts us from our first priority of knowing God. When we don't observe these priorities, nothing seems to work for us. Thomas à Kempis tells why: "The reason why many things displease you, and often disturb you, is this, that you are not yet perfectly dead to yourself, not detached from all earthly things" (Thomas à Kempis, *Imitation of Christ Selections,* Wheaton, Ill.: Tyndale House Publishers, 1969, p. 41). One of those earthly things is our quest for happiness.

Just when I think I've learned this, I see that I have more to learn. When we planned to move to a nearby town, my friend Rosemary, whom I consider a prayer giant, prayed that our family's move would be smooth. Since our house sold immediately, she prayed that we would find another house quickly and that the seller would be agreeable to our offer.

We began looking and it was terrible. I have trouble making decisions, so I usually let my husband Greg pick a house and I adjust. This time he wavered for 10 days among seven houses. Each day he changed his mind. He hated—even detested—each of the three I liked. For a week, I had a headache and couldn't sleep. We finally made three offers which were all turned down.

I'd had it. Hadn't Rosemary prayed that the move would be smooth? We finally compromised on a house that neither of us liked all that much and difficulties multiplied.

I went to the Lord in prayer. What about Rosemary's prayer? Why couldn't I sleep? Why did my head hurt? Why was God allowing this to be so difficult? Couldn't He intervene and provide the "perfect house" if He wanted to?

I thought of how I resembled Habakkuk. Like him, I figuratively stood on the wall, crossed my arms across my chest, and said, "Why, God?"

But I also differed from Habakkuk. He argued for the cause of justice because God's people weren't glorifying His name. I was only arguing for a softer, easier life. Habakkuk cared so much for God's causes, for God's name! I cared mostly for my own comfortability. And, I seemed to want a guarantee of continued future happiness for the rest of my life.

That "one day at a time" phrase came into my head and I prayed, "Thank You that we now have a house. Thank You that we can even afford one. Help us to use whatever house we live in to Your glory." Then I paraphrased Habakkuk's resolution in 3:17-18, "Though the house I shall live in has not yet appeared and I know not my moving date, I will rejoice in the Lord."

A few minutes later the irritation within me rose again. I gritted my teeth and spat out my prayer: "I will rejoice in You, Lord. Please answer my needs. I trust You know them better than I do."

For too long in my life, I've tried to use God as my shortcut to happiness. I have loved His blessings more than the Blesser Himself. When I use Him this way, I am always disappointed. When I come to Him as His child seeking His mercy, He never fails.

Feelings as Spiritual Thermometers

During a youth meeting several years ago, I handed out blank graph paper to a group of college kids and asked them to chart their spiritual growth over the last year. I looked over their shoulders as they worked. Debbie marked the day she moved out on her own as a spiritual low point. I asked her why.

"Because it was so depressing to be alone," she answered.

"But was God working in your life?" I asked.

"Yes," she said, "I'd moved out on my own. It was hard, but it was great. I felt like God was protecting me."

Then Debbie tipped her head and asked, "You mean spiritual highs and emotional highs are different?"

That thought hit me at the same time. Spiritual growth hurts. Our worst times emotionally may be some of our best times with God. We may feel far from God, but we aren't far away at all. That stretching hurts and we're depending on God more than usual.

Too many times I'm looking for good feelings, especially feelings of closeness. I discovered myself doing this once during a worship service. Throughout our customary 30 minutes of singing and worship, I never felt God's presence. I was feeling unusually blue, so I concentrated hard and thought, *I'm seeking You, Lord.*

Finally, in almost a disgusted sigh to myself and a half smile, I remembered, "Lo, I am with you always" (Matthew 28:20, KJV). I realized that God lives in me. What was I straining so hard for? True, I hadn't felt close to God, but He lives inside me. There's no question about this. The truth is that God is seeking me. He loves me.

Since that day, I begin the worship service by saying to God, "Thank You for living inside me, for being here in this service. I'm planning to worship You. I hope You enjoy it."

Spiritual growth can be a struggle. Heroes of the faith such as Abraham, Moses, and Paul struggled not only with God's people but with God's will in their lives.

In the spiritual life, pain isn't always a negative feeling. We get comfortable in our ruts and we like them. Changing anything about our life is inconvenient. Think about the pain necessary to go on a diet. You have to be terribly uncomfortable to change—your clothes don't fit, you can't get up as easily, people make unkind remarks. When we finally feel uncomfortable enough, we change. We dislike change so much that we only do it when it's less painful to change than not to.

That's where pain helps us in our spiritual growth. After repeated moments of weeping and sadness, then we consider change. What if I tried being grateful instead of feeling sorry for myself? What if I didn't speak out in anger? What if I offered some kind words to my exasperating teenager?

If we think it will help us stop feeling miserable, we're willing to try thankfulness, meekness, kindness.

Switch the Questions

I used to ask myself, "Am I happy?" "Am I having a good day?" If I were having a good day, it would be easier to be nice to people. Life is such that the answers to these questions were often no. It wasn't easy to be nice to my husband, my children, my co-workers. It was impossible to be nice to the forgetful clerk, the incompetent teacher, the critical women's group leader. They all committed the cardinal sin—they made me unhappy.

Perhaps one of the most miserable moments is running errands. My children are old enough to have better things to do, but not old enough to be left at home alone. They have to go with me and they complain about it.

After settling an inevitable squabble, I used to ask myself, "How do I feel?" "Why am I so unhappy?" The answers were not pleasant.

As I've grown in the Lord, I've substituted those ill-fated questions with these:

Am I practicing the presence of God?

Am I fulfilling His purpose in my life—bringing the people I know into maturity? (Colossians 1:28)

Am I praising God for who He is?

When I try to answer these questions appropriately, the tone of things in the car changes. I immediately pray and thank God for being present with me. I try to think of ways to minister to my kids (we talk about the Bible, I ask them about their day). Without even intending to, my prompting distracts them from their squabble. I flip in the cassette tape of praise songs and I hear my daughter's voice from the back seat join in.

It's tempting to say, "Gee, I want to feel happy so I'll focus on God." Somehow it doesn't work. You know you're manipulating your spirit and the sincere desire for God isn't there.

I have to focus on God without desiring happiness. And that's best anyway. The more I know Him, the more I want to focus on Him.

FOLLOWING DOCTOR'S ORDERS

Plot yourself on the continuum below. When you're going through trials, does your position on the continuum change? If so, mark it. What changes your position on the continuum?

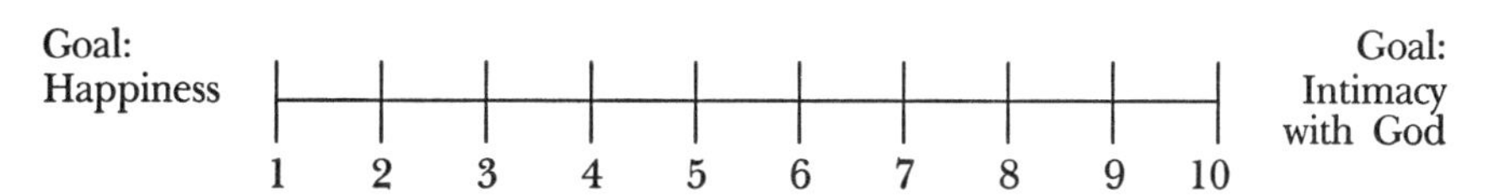

1 = quest for happiness
5 = quest for happiness and intimacy with God
10 = quest for intimacy with God regardless of happiness or comfort

Look at the items below. They are "needs" that humans are supposed to typically satisfy. Which ones drive you to seek happiness more than intimacy with God? Place a check mark next to them.

☐ need for material things to make you happy
☐ need for close relationships to make you happy
☐ need to feel good

Make a list of your favorite Scripture verses. How many of them involve the blessings you hope to receive from God? Pick out some verses that emphasize your love for God to add to your list of favorites (see Psalm 18:1; 43:4; 116:1-2).

Go through some Psalms such as 142. Look for signs that the writer sought God himself. Sometimes he also sought the blessings of peace or security or asked for vengeance. Read a praise Psalm such as 149. In what way is praising God seeking Him? Try it and see if it seems the same to you.

CHAPTER SIX

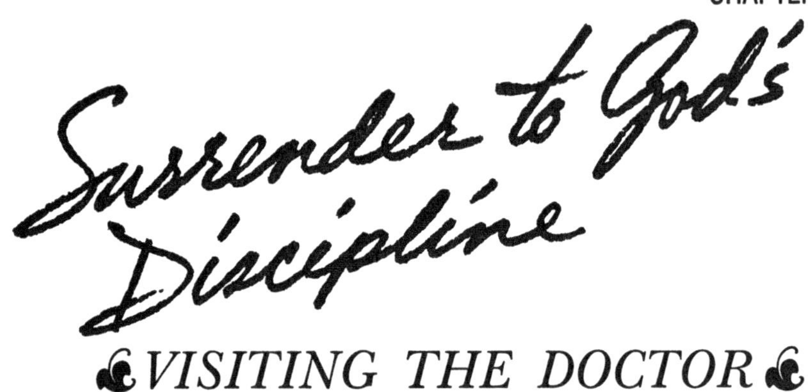

VISITING THE DOCTOR

"God works in mysterious ways," is one of those overused clichés. Yet it's worn out because it's true. It fits many situations that we don't understand.

Most of us spend lots of energy guessing what God's purpose might be in perplexing situations. We want to understand, we want to comfort ourselves. Ultimately, the greatest comfort is that God is a faithful, loving, just God. He helps us stand firm.

Help us, Lord, to surrender ourselves to You.

Read Hebrews 12:12.

1. *Therefore* is another way of saying, "Having said all of this." Summarize in a sentence or two what has been said so far in verses 1-11.

2. Arms and knees were important in battles of old. Arms held weapons and knees were what kept people moving forward. What do we use to progress in the Lord?

3. The command in verse 12 is the opposite of "losing heart" (see verses 3, 5b). Describe what it's like to lose heart. How do you feel? What color is your life when you lose heart?

   How do you currently fight the battle of losing heart?

Read Job 1:1–3:1.

4. What limits did God put on Satan? (1:12; 2:6)

5. What atypical response did Job offer God after he suffered such great losses? (1:20-22)

   What are the subtle ways we "charge God with wrongdoing," which Job did not do? (1:22)

   Is asking, "Why me, God?" charging God with wrongdoing?

6. When people discourage you as Job's wife discouraged him, how do you usually react to them? (2:9-11)

Read Job 32:1-3.

7. Why did the men stop answering Job?

   Was Job right to think that he was righteous? (See Job 1:8, 2:3 for God's opinion of Job.)

8. Why was Elihu angry with Job? With the three friends?

9. When you watch others hurt, do you usually respond like:
   - ☐ Job's wife who talked negatively about God
   - ☐ Job's three friends who tried to figure out why this was happening
   - ☐ Elihu who reaffirmed God's justice and mercy (36:2ff) and who confronted Job in his self-righteousness (33:12ff)

Read Job 38:2.

10. What was God's complaint against Job?

11. What purpose do you think God's questions and comments in 38:4–41:34 served?

Read Job 42:1-6, 10.

12. What did Job repent of? Why?

    Did Job repent because God prospered him?

    When have you been guilty of speaking about things you did not understand, things too wonderful for you to know? (v. 3b)

    What do you think is the wisest thing to do when you experience trials and don't understand God's reasons?

13. What promises do we have from God that help us trust Him even when we don't understand?

    Psalm 33:18

    1 Corinthians 10:13

    Isaiah 61:10

John 14:16-17, 15:4

2 Peter 3:9-10

Would you say that Job "lost heart"? Job obviously grew weary (v. 3), but he didn't lose faith in God's love.

Read Hebrews 12:13.

14. How do we take "level paths"? Search the verses before and after Proverbs 4:26, from which this phrase is quoted.

    In what way will "level paths" of righteousness help a lame person prevent complete disablement?

15. Who caused the "lameness" in these situations?

    David's life—Psalm 51:10

    Judah's defeat—Jeremiah 50:17

    Judah's defeat—Lamentations 3:4

    How did God respond when someone else afflicted Judah? (Isaiah 35:3-5)

    What was God's response when He afflicted Judah? (Micah 4:6-7)

Why does God purposely disable someone? Consider David and the background of Psalm 51; Judah and her idolatrous life before captivity.

16. Here are some practical ways to surrender to God. Write one specific area next to each.

    Forsaking lifelong bad habits

    Considering His will in all things

    Accepting difficulties without questioning God or justifying myself

17. Underline one or two of the ways you listed that you need to surrender to God.

BINDING UP WOUNDS

As the car whirled to a stop, I tried to figure out what happened. My husband's head was bleeding. I couldn't move because my leg was trapped. I finally realized I could move my arms. I raised the palm of my right hand in front of my face. I could see that my hand was almost a full inch to the left of my wrist. I recognized the injury from playgrounds in my childhood. I had a broken wrist.

Ten weeks and two casts later, I watched as the doctor sawed off my cast. I held up my scrawny arm in front of my face the same way. A miracle had occurred. The hand was positioned exactly on top of the wrist.

This "miracle" occurred because a surgeon had stuck a stainless steel pin through my hand and another through my elbow. He had moved my hand over to realign it with my wrist and then held it in traction with the pins and a huge cast. Eight years later, the only remaining evidence are two half-inch scars from the puncture wounds on my right hand. When people ask about the scars, they cringe at the thought of a metal pin being shoved sideways through my hand.

I don't. I know that I was anesthetized during the procedure and given extensive medication for days afterward. Besides, I remember seeing the disfigured wrist—that's what makes me cringe. I've even imagined what it would be like if the surgery hadn't occurred. What if I tried to cook or type or play the piano today with a disfigured wrist?

When I moan and groan about the pain of spiritual growth, I wonder if my spirit looks as disfigured as my wrist would have looked without surgery. When I look at it that way, I'm glad for the trials I endured. They're keeping my spirit from being disfigured.

The problem is that we don't have powerful anesthetics for spiritual pain. We go through them wide awake. So we fight them, deny their existence, even pretend they don't hurt. We need these trials, but we can barely stand them.

Sometimes endurance seems too tough. Strengthening feeble arms and weak knees seems beyond our ability. We long for rugged spiritual endurance so we can go the distance with God.

How do we do that?

One way is changing how we answer anxious questions like "Why me?" and "When will it be over?" Little by little, we give ourselves the right answers each time.

Here are some "right answers" about the Christian life.

Expect Trials, Expect God

Trials stun us. Somewhere we got the idea that Christians have special protection from trouble. Sometimes our, "Why me?" questions suppose that a certain trial should happen to the wicked guy down the street, but not me.

Hudson Taylor, a missionary who led thousands of souls to Christ in China and recruited many others to do the same, experienced terrific trials. On the mission field, he lost his daughter, his wife, and even his own health (Marshall Broomhall, *Hudson Taylor, The Man Who Believed God,* Edinburgh: R & R. Clark, 1929, pp. 97, 143, 153). No one is exempt from trials; frequency of trials is not a sign of spiritual weakness or second class Christianity.

Peter even told the brave, early Christians: "Do not be surprised at the painful trials you are suffering, as though something strange was happening to you. But rejoice that you participate in the sufferings of Christ, so that you may be overjoyed when His glory is revealed" (1 Peter 4:12-13). We not only expect the trials, we expect to see what God will do with them. We view them as opportunities for God to work.

On a certain Tuesday a few months ago, two major crises in my life were resolved. I'd leaned heavily on God through both, but I was relieved they were over. I was euphoric to see them end on the same day. That night, my husband Greg found out he was being laid off from his second job. When he told me, my first reaction was to panic because that income was helping us get by. Then I thought of my recent trials and how faithful God had been. Instead, I prayed: *God, You are so great. Cash is no problem for You. I wonder what You'll do this time.*

Trust That God Has a Reason

We rarely understand the reason for our trials immediately. My friend Annette didn't understand a major disappointment for two years. We both knew a well respected pastor whose theology on a certain point fit hers neatly and was rather unusual. She wrote to him asking if they could write a book together. He agreed.

As it worked out, Annette was forced to go back to teaching and this meant they couldn't write the book together. She was disappointed and referred with regret to this opportunity as her "big chance that got away." She questioned why it worked that way. "Maybe we're off track and this theological point is wrong," she pondered.

Two years later, this pastor got a divorce. The wife explained to us that he had been having affairs with younger women (our age) for many years.

Annette was stunned. "I had always been attracted to him," she confessed to me. He was older and smarter and wiser and she adored him. At the time they talked about writing the book, Annette's husband was going through a mid-life crisis. Annette was supporting him emotionally, but he ignored her a lot.

"What if I had started spending hours with this man?" Annette cried. "I wish that I could tell you that nothing would have happened, but more righteous people than I have fallen. I believe now that God spared me from temptation that was too steep for me."

We don't know, of course, if that's what God had in mind, but now I wonder when I go through unpleasant situations. Is God intervening and protecting me from dangerous situations? Am I short-sighted to complain about my "big chances that got away"?

Accept That You Don't Understand

As we grow, we understand more about God, but it takes even more time and growth to understand how God's ways are worked out in our lives. "A man's steps are directed by the Lord. How then can anyone understand his own way?" (Proverbs 20:24) The path of our lives perplexes us—we don't understand how God's ways are at work in us.

My friend Mary describes her Christian life this way: "It's as if I'm riding in a chariot with Jesus. I'm uncomfortable because He's driving the chariot too fast. The wind is blowing in my face and my hair is streaming straight back. Jesus and I are both straining forward. I have no idea what's going on, but I know it will be OK because Jesus is driving the chariot. I've considered jumping off, but that would be even more scary."

Don't feel unspiritual if your Christian life perplexes, even confuses you. God certainly isn't the author of confusion (1 Corinthians 14:33, KJV), but as we humans interpret His ways, we often get confused. That's partly because we keep second guessing God. Since His ways are so different from ours, we don't guess well.

Like Annette, I decide that a certain trial occurred to teach me a certain character quality. Then everything changes and it looks as if God didn't intend that at all. Besides, my view of God gets out of balance. Some days, I assume that God wants to solve all my problems; other days, I practically accuse Him of creating them all.

Being perplexed by God isn't a sign that you're an inferior Christian. You're actually in good company. Paul wrote, "We are hard pressed on every side, but not crushed; perplexed, but not in despair" (2 Corinthians 4:8).

Why is it so normal to be baffled by God? In *The Last Battle,* the last king of Narnia describes the lion, Aslan, who represents Christ, this way: "[Aslan] is not a tame lion" (C.S. Lewis, Chronicles of Narnia, *The Last Battle,* New York, NY: Collier Books, 1976, p. 16.). God isn't "tame" either; He doesn't do as He's told. He doesn't meet our deadlines. He defines love in ways that stretch beyond our horizons.

God has a perfect will that's far beyond our understanding. There are incidents in our lives that we may not understand for years, that we may never understand.

Not being able to understand things is foreign to our culture. I used to continually quiz Dick, my "computer man," as he upgraded and repaired my computer. I was curious because my computer is such an important tool to me. I finally realized that I never understood his answers. Occasionally when I did understand what he said, I couldn't remember it the next day.

Now, I've decided not to ask him questions. I trust Dick. I tell him my problem and he repairs it or invents a program to help me. I smile, pay the check, and even send him a thank you card later.

I've given up my need to understand everything that's valuable to me; I trust a person instead. I'm trying to do the same thing with God as He works on my life.

How do we get by when we don't understand? When the Apostle Paul talked about being perplexed, he also said, "But we have this treasure in jars of clay to show that this all-surpassing power is from God and not from us" (2 Corinthians 4:7). Paul had surrendered himself over to God.

Surrender Over and Over

As harsh as it sounds, it doesn't matter if we understand. God works His mighty ways whether we understand Him or not. His miracles don't have to be understood to be wonderful. If you search the Book of Job, you won't find the reason why God allowed Job to suffer.

Having the answers doesn't always make them easier to accept anyway. Look what happened when God explained His plans to Habakkuk. Habakkuk was appalled that God would use the heathen Babylonians to teach Israel a lesson. Then when God went on to explain that He would punish Babylonia too, Habakkuk seemed somewhat satisfied. Habakkuk's finest moment was when he surrendered to God: "Yet, I will rejoice in the Lord" (3:18).

Most of us need to turn in our detective badge on God's ways. I long for the day when I don't have to come to the end of my wits before I admit that:

—I can't change what's happening.

—Only God can change it.

—I'm willing to wait and watch Him do whatever He wishes.

Oswald Chambers described a Christian as "not one who proclaims the Gospel merely, but one who becomes broken bread and poured out wine in the hands of Jesus Christ for other lives" (Oswald Chambers, *My Utmost for His Highest,* Westwood, NJ: Barbour and Company, Inc., 1963, p. 40). In other words, we have to be willing to be crumbled in His hands and poured down the drain for Christ.

To be honest, it scares me to think that way. I'm surrendering myself to a Savior who faced His own murder with openness to God's grace.

Yet to refuse to surrender is even more painful. To resist God would have been like tensing up as the stainless steel pins were inserted through my hand. I'd rather go limp.

So when things get harried, I often cover my face with my hands and bow my head. I pray, "Lord, I give up." I think of God's love for me and for a minute it seems as if I am burying my head in the mane of Aslan (the lion representing Christ in the Chronicles of Narnia) as a small child buries his head in his mother's skirt. It isn't a desperate prayer, but a quiet one. For the moment, God has me where He wants me: surrendered.

FOLLOWING DOCTOR'S ORDERS

Write a note to someone who's going through a trial. Express words of comfort that show you are sure of God even though you aren't sure of what He's doing.

Make a list below of all the things you don't understand (how a child's body grows, how your computer works, how your television and telephone work). Thank God that He understands these things. Name issues or circumstances in your life and tell God that you don't understand them either.

Read chapter 2, "The Rashness of the King," of *The Last Battle* of the Chronicles of Narnia. How did it comfort King Tirian and Jewel to remember that the old stories said that Aslan was not a tame lion? How did these two use that fact to assume the worst of Aslan a few pages later?

Think of Christians you know who seem surrendered to God. What attitudes make you think so? If possible, talk with them about this subject and ask these questions:

- ☐ What experience was the most helpful in teaching you to surrender to God?
- ☐ What do you do when you don't feel like giving in to God?
- ☐ What benefits do you receive from giving in to God?

CHAPTER SEVEN

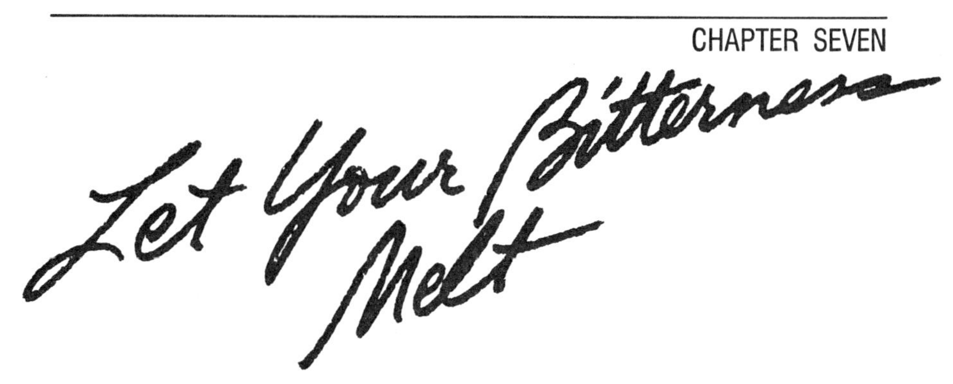

VISITING THE DOCTOR

What happens when we feel that the discipline of the Lord is too much to bear? We're not sure whether God initiated these disciplinary acts or they are mere circumstances through which we can learn the Lord's discipline, but we don't like them. We wonder why God allowed them.

Even mature Christians can become bitter. This bitterness wears different faces. Some people get outright angry with God and talk about it. Others feel frustrated and begin ignoring their spiritual lives.

When we do this, we miss the grace of God; that is, we reject God because we think He has rejected us.

Help us, Lord, to be honest about our past hurts. Show us how to set them aside and reach out to You.

Read Hebrews 12:14.

1. How does a person manage to "live in peace with all men" and still "be holy"?

2. What would you say to someone who thinks that "without holiness no one will see the Lord" means you have to be nearly perfect to be saved?

   What is one way that holiness grows within a person? (See verse 10.)

Read Hebrews 12:15-17.

3. How does Ephesians 2:4-5 define *grace*?

   Why are those who are experiencing the discipline of the Lord in particular danger of "missing the grace of God"?

4. Micah 4:6 creates a picture of God extending His grace to those He disciplined or "brought to grief." Why do people miss God's deliberate offer of grace, especially after He's "brought them to grief"?

5. What wrong attitude is involved in "missing the grace of God"? (James 4:6, 10) Does that make sense to you? Why or why not?

6. Galatians 5:4 says that those who justify themselves by the Law fall away from grace. Why does self-justification separate us from God?

7. What kinds of experiences create "bitter roots" in women's lives? Name some specific harsh or unpleasant experiences that leave a bad taste in your mouth or that produce feelings of grief, disappointment, or resentfulness.

   How can one person's bitterness affect other people? Has another person's bitterness ever affected you?

8. How is Esau described in verses 16-17?

Read Genesis 25:29-34; 27:30-41.

9. What reason did Esau have for selling his birthright?

10. Why do you think the writer of Genesis said that Esau "despised" his birthright? (v. 34)

11. What was Jacob's part in stealing Esau's blessing? (27:1-19)

12. In Genesis 27:36, Esau said that Jacob deceived him twice. Was Jacob deceitful both times? Why do you think Esau's view was distorted?

13. Hebrews 12:17b implies that Esau was bitter about something else that happened after he lost his blessing. What was it? Could Isaac have done anything to help Esau? How did Isaac seem to feel about what happened? (see 27:33, 35)

14. How do you think pride affected Esau? How could he have responded correctly to God's discipline? (He planned to kill Jacob [v. 41].)

15. God obviously chose to bless and use Jacob, as deceitful as he was, and Esau suffered for it. When have you been passed over and resented the choice?

Read 2 Chronicles 16:1.

16. To introduce yourself to King Asa of Judah, skim 2 Chronicles 14:1-6 and 15:1-7. How would you describe Asa?

Here are three people to remember:

- ☐ Ben-Hadad, King of Aram (enemies of God's people, generally wicked)
- ☐ Baasha, King of Israel (the northern 10 tribes which had turned mostly to other gods, but were still God's chosen people)
- ☐ Asa, King of Judah (the more faithful southern kingdom, whom God protected)

Read 2 Chronicles 16:2-6.
17. What did King Asa do in this passage?

Read 2 Chronicles 16:7-14.
18. Why was God displeased with Asa?

What had been God's goal for Asa—and for us? (v. 9)

God disciplined Asa for this by keeping him continually at war. Why could constant war have been good for Asa? How could it have helped him achieve what he violated by leaning on the King of Aram?

How did Asa respond to God? (v. 10) What part did Asa's pride seem to play? Would you say that Asa "missed the grace of God?" (Hebrews 12:15)

Perhaps Asa's illness (v. 12) was further discipline of the Lord. How did Asa respond to it?

When have you rebelled against God's discipline as Asa did? How do you show it? For example, do you get angry as Asa did? Do you withdraw from God?

19. The discipline of the Lord results in sharing God's holiness, which is exemplified by righteousness and peace (Hebrews 12:10-11). Reread this passage and find evidence that Asa did not seek righteousness or peace.

BINDING UP WOUNDS

Even "nice girls"—women who attend church, read their Bibles, and pray—can be devoured by bitterness. For many years, whenever I didn't agree with God's will, I'd question God and then sulk for a while. I was a lot like the typical teenager who runs to her room, slams the door, and doesn't come out for a while. Sometimes out of pride, other times out of stubbornness, I refused to submit to God's will. I wanted things done my way, or I'd quit playing the game. I still went to church, I listened to sermons, and I even served faithfully, but my heart wasn't in it. My personal Bible reading and prayer life languished.

King Asa of the Old Testament was a good guy, one the best of Judah's kings. Yet he tripped over God's discipline. He didn't trust God for victory in battle and the seer confronted him: "The eyes of the Lord range through the earth to strengthen those whose hearts are fully committed to Him" (2 Chronicles 16:9). This discipline was meant to train Asa, to help him become fully committed.

How different the last years of Asa's life would have been if he would have submitted to God's discipline. Instead Asa got angry and took it out on the people. When he contracted the foot disease, he could have turned to God, but he didn't. It looks as if God was trying to get Asa's attention, but to no avail.

Bitterness: What Is It?

Bitterness is that lasting anguish that results from times when we're angered, hurt, or disappointed. We resent these experiences and refuse to move on.

The root word for *bitterness* means "to cut, to prick" (W.E. Vine, Merrill F. Unger, and William White, Jr., *Expository Dictionary of Biblical Words,* Nashville, TN: Thomas Nelson Publishers, 1983, p. 68). God reserves the right to inflict these wounds or to allow them to be inflicted. The psalmist even spoke of the bones God had crushed (Psalm 51:8b).

These wounds from God can be compared to surgical wounds. They're carefully planned to do us the most good.

When we're bitter, we feel wounded or knifed by God. That's because surgical incisions are little more than strategically placed stab wounds. They hurt just as much even though they're intended for a noble purpose.

We feel betrayed because we misunderstand the incisions of His discipline as stab wounds. We feel that He intended to hurt us and we decide to hurt Him back. It's our way of being prideful, of saying, "I don't have to take this."

I'm sure Stephanie, who used to be the junior youth group leader, felt that way. She faithfully served up wonderful junior youth group meetings, but she didn't always attend worship services, and she never attended anything else.

I could tell that she knew a lot about the Bible and I found out that she had once been the Bible school superintendent—and a good one. But she had tried to initiate changes that weren't always appreciated. It was sad because they sounded like changes that would have helped our church.

As I got to know her, I asked her about her church involvement. She said, "I worked really hard to help this church and they didn't care. Now I do my own thing and that's all I do."

I asked her how her faith was holding up. She said, somewhat defensively, "I still believe in God. Don't worry about me."

As other hurts came into her life over the next few years—her brother died, her job changed—Stephanie used her stiff upper lip technique. Her attitude toward life said, "I'm tough. I can take it. I don't expect anyone to be nice to me."

Stephanie no longer attends church now. I look back and wonder what she would have said if I had asked her, "You believe in God, but do you still trust Him?"

Bitterness Never Sits Still

It's contagious. As Stephanie quit participating in church, so did her relatives. They took up an offense for her against the church leaders and the poison of her bitterness spread.

Bitterness is also shadowy in its growth. Susan thought she'd accepted her mother's death, but she quit volunteering for anything at church. When someone at church made an unkind comment, she quit coming to church regularly. She now realizes that she was mad all along about her mother's death, but she couldn't talk about it.

The writer of Hebrews paints a vivid picture of the bitterness process:

"See to it that no bitter root grows up to cause trouble," (Hebrews 12:15). "The seed, the root, lies hidden and reveals its power slowly" (B.F. Wescott, *The Epistle to the Hebrews,* Grand Rapids, MI: Eerdmans Publishing Company, 1977, p. 407).

As each potentially embittering situation occurs, God gives us the grace we need to handle it. We may "miss" His grace by refusing to accept it and choosing to be angry. Our view becomes distorted as Esau's was. He willingly sold his birthright out of his own godlessness, but later said that Jacob stole it (Genesis 27:36). In his bitterness, Esau tried to blame both losses on Jacob. In this way, he missed the grace of God.

God's Wonderful and Terrible Grace

"Missing the grace of God" is not like missing a bus or failing to see a comet whiz by. God actively gathers and assembles those He's brought to grief (Micah 4:6). You've got to go out of your way to miss God's grace. God offers opportunities to be comforted, to be restored, to be counseled, and we openly ignore them, reject them, or put them off. We form our own rut of rejecting God and sit in it.

Grace is God's way of "cutting us some slack." He forgives us or lets us off the hook even before we ask for it. The moment we even think about approaching God, He's standing with outstretched arms saying, "I'm ready to love you." We need God's grace because it serves as our umbilical cord to God—it helps us find our way back to Him when we grow bitter.

When you're feeling bitter toward someone, it's a terrible burden to have that person love you anyway. She should be mad at you, but she waits on you instead. You know that the harmony of the relationship is being delayed because of you and your stubbornness.

In the same way, God offers us His grace and forgiveness at every turn, and we have to blind ourselves to ignore it. As He sends unexpected blessings and undeserved love our way, we have to steel ourselves against wanting to be right with Him. It's as if He keeps following after us, waiting patiently until we respond. As we grow in the Lord, we mature and surrender to Him sooner each time.

This requirement of surrender explains why this passage seems so hard on Esau. Have you wondered why it refers to him as, "godless," while Jacob the Schemer is listed in the Hebrews 11 "Hall of Fame of Faith"?

It's true that Jacob was a man of mixed motives, but he grew into a man of deep faith. The more troubles he had (standing up to Laban, coming home to Esau), the more he listened to God (Genesis 31-35). When Esau lost the birthright and the blessing, he became bitter. The difference between Jacob and Esau is that Jacob was open to God's discipline and Esau was not. That difference allowed Jacob to be the raw material for God to work with.

Trials: Use Them or Lose Them

Jacob, who was one of the most deceitful men God ever used, required lots of discipline from the Lord. Just as Jacob had swindled Esau, Laban (his father-in-law) swindled him! Jacob found himself having to settle accounts with Laban and then face Esau. The night before Jacob met Esau, he wrestled with "a man" (God) until he "won" (Genesis 32:22-32). Jacob didn't "win" by beating God, but by refusing to stop short of gaining a blessing. He also gained a limp—his mark from God.

I've been good at wrestling with God myself. Only I wrestle until I'm tired, not until I've gained a blessing. When my husband was in the ministry, I seemed to pick up a new grievance every few years. I was mad because we were underpaid at one point; at another time, a staff member who was threatened by him criticized him; still another time his job description was changed drastically after we'd been on staff for a few months. Now that I'm older and wiser, I see that these things aren't that unusual in the ministry or any other job.

I remember sitting at a women's retreat, just a "regular church member" for the first time in my adult life. I carried all my bitternesses with me. I can't remember what the speaker said, only that she quoted Jeremiah 29:11: "For I know the plans I have for you, says the Lord, plans for welfare and not for evil, to give you a future and a hope."

I sat in the chapel, staring out the huge glass at the evergreens in front of me and wept. I was angry at people in the churches we'd served; I was angry at God for not allowing my talented husband to "succeed" the way he wanted to.

The last word in Jeremiah 29:11, *hope,* reverberated in my ears. Was there hope for us? Hope for me? I had buried my bitternesses over many years and I was finally ready to get rid of them. I was ready to surrender to God all the situations I hadn't understood, all the people that confused me. I'd spent many years wrestling and I was ready for the blessing. I was ready to accept God's grace, to go to Him for love and comfort.

Also like Jacob, I would forever walk with a limp. But my limp shows up in surrender, not arrogance. My limp is not a mark of God's anger, but of His grace.

FOLLOWING DOCTOR'S ORDERS

Sit in a quiet spot and meditate on Hebrews 12:14-17. Then think of a person or circumstance about which you feel bitter. Turn the palms of your hands up to release it to God. Then turn your palms down and receive the grace, surrender, and hope God wants to give you. Do this with as many things as necessary.

Make a list of events in your life that have assured you that God loves you. Make another list of sins for which you have been forgiven.

Read Psalm 13. The writer sounds as if he could have become bitter, but didn't. Look through the Psalm and find phrases that show how he surrendered to God's grace. If you are feeling bitterness toward God, paraphrase this Psalm and insert your situations and feelings.

Look at the words to a song such as "He Giveth More Grace" or "Something Beautiful." How have they been true in your life? How have you seen God grant you special favors to help you overcome potentially embittering experiences?

CHAPTER EIGHT

Watch His Discipline Help You

VISITING THE DOCTOR

We hear a lot about Christians being "diamonds in the rough." Our trials, temptations, and discipline of the Lord smooth out our rough edges and pressure us into being beautiful gems for God.

What will that polished glow look like? What changes can we expect to see in our lives because God has either corrected us specifically or because He has allowed trials in our lives?

Father, we thank You for the discipline that we've experienced. Help us to accept it further and surrender to You. Thank You, that we know You and love You better, as a result.

Read Hebrews 12:10-11.

1. What spiritual qualities result from God's discipline?

"Share in His Holiness."

2. To be "holy" means to be separated or set apart. In the Old Testament especially, it meant being "pure" or "devoted" (W.E. Vine, Merrill F. Unger, and William White, Jr., *Expository Dictionary of Biblical Words,* Thomas Nelson Publishers, p. 113.) How does being disciplined by the Lord make you feel different or set apart? Did you feel purified or especially devoted to God? Give an example from your life, if possible.

On the left side of the line below, list the characteristics of holiness (or sanctification) from 1 Thessalonians 4:3-8.

How do you think God's discipline helps us share in each of the above aspects of holiness? Write your thoughts on the right side of the line above. Then give an example from your life, if possible.

3. Read the following passages and search for a phrase similar to "share in His holiness." Write that phrase below the reference. Then write the result of the phrase similar to "sharing in His holiness." The first one is done for you.

   Galatians 2:20
   phrase: crucified with Christ
   result: Christ lives in me, I live my life by faith

   Philippians 3:10
   phrase:
   result:

   1 Peter 4:13
   phrase:
   result:

   2 Peter 1:4
   phrase:
   result:

4. How do these phrases help you better understand what "sharing in His holiness" probably means?

5. How does identifying with the sufferings of Christ help you grow spiritually? Give an example from your life. For example, have you ever come through a trial feeling closer or more loyal to Christ than ever? Have your trials forced you to seek God in a deeper way?

Read Psalm 57:1-11.

6. Why was David so afraid of Saul? (1 Samuel 19:1, 9-12; 20:30-33)

7. How would you have felt if you were David? He was anointed to be king and the present king was wicked and wanted to kill him.

   Pick one verse from Psalm 57 that would describe your feelings.

8. Even though David had been fleeing from Saul for some time, Psalm 57 doesn't reveal that David felt bitter. Listed below are some of the feelings he did express. Write the verse references that point out these qualities.

   - ☐ dependence on God
   - ☐ loved by God
   - ☐ praise for God
   - ☐ faithfulness to God
   - ☐ horror at his own predicament

   Which of the above reactions have you shown when you've been disciplined by God specifically or He has allowed life's circumstances to discipline you? (Note especially the first four.)

9. How does David's psalm illustrate the phrase, "rejoice that you participate in the sufferings of Christ"? (1 Peter 4:13)

Read 1 Samuel 24:1-7.

10. Comment on David's situation at this time. (Either this passage or 1 Samuel 22:1-5 is the background for Psalm 57. Even scholars aren't certain.)

    How did David "share in God's holiness" by showing righteousness and peace in this situation?

11. What about David's life tells us that ordinary, sinful people can "share in His holiness" too at times? (See 2 Samuel 11 if you need help.)

Righteousness and Peace

12. Righteousness means right living, that is, living faithfully and truthfully. How does the the discipline of the Lord equip a person to live right?

    Have you "survived" a trial with a greater determination to do right? Do you have any strong commitments to honesty, tenderness, pure sexuality, because of trials you've suffered and survived? Explain.

    Place a check mark by any of the *results* in question 3 that you think deal with righteousness.

13. Peace is more than the absence of cnaos. It's that deep sense of rest and contentment with God, self and others. Why do you think trials produce this peace? How has this kind of peace come about in your life?

    Place a check mark by any of the *results* in question 3 that you think deal with peace.

BINDING UP WOUNDS

We recently changed dentists. This new one is a lot of fun. We even joked during the exam. But when it was time to come back to have a tooth filled, I didn't want to go. I had to force myself to have it done. No matter how much fun a dentist is, I only go because it's good for me.

Most Christians feel the same way about the discipline of the Lord. It's not something you outright choose. Even though the Bible says to rejoice in sufferings, I know of few Christians who have advanced that far in their spiritual growth.

But the Bible puts that goal out there in front of us. Frankly, it's smart. Accepting God's discipline keeps us from wasting energy resenting or questioning Him. We're finally cooperating with Him and our spiritual life can zoom ahead.

So, let's figure out some of the unexpected advantages of the "discipline of the Lord" and motivate ourselves to go along with God next time.

"Sharing" His Holiness

The phrase, "share in His holiness," is a mysterious one that deserves our exploration. This "participation in God's eternal nature" (2 Peter 1:4) is a cross-breeding of sorts. We, of the human nature that seems bent toward hatred and crime, become sharers in the divine nature by God's grace.

W.E. Vine describes holiness as a "state predetermined by God for believers, into which in grace He calls them, and in which they begin their Christian course and so pursue it" (W.E. Vine, Merrill F. Unger, and William White, Jr., *Expository Dictionary of Biblical Words,* Nashville, Tenn.: Thomas Nelson Publishers, 1985, p. 307). It's like being born into the British royal family. You receive special treatment, but it takes you a long time to live up to it.

Part of that specialness is when we begin to identify with Christ and

His sufferings. It builds an intimacy with Christ. It's something like the camaraderie that war buddies develop in battle, except that Christ is like the army general and we're the young recruits peeking over the foxhole. We appreciate Christ's sacrifice on the cross even more because we've tasted His agony close up.

The first time I remember experiencing this was years ago in college. I was accused of cheating because what I said had been misinterpreted. I defended myself and corrected the misunderstanding, but the professor didn't believe me. I took it to the higher-ups, but they sided with the professor.

I remember sitting on the bed in my dormitory room crying for a long time. I prayed, "Lord, this really hurts to be wrongfully accused. I'm so embarrassed that people think this of me. I'm sorry my sin put You on the cross and everybody thought so badly of You. You must have loved me a lot to have died for me. Thank You."

Most of the time I suffer because I'm wrong, but this time I was innocent. This small similarity to Christ's experience built a strong bond to Christ within. Now I try to use every experience in which I'm wrongfully accused to bond to Him even more. As these discipline experiences continue, we learn to constantly seek God. We pray, we read our Bibles, we eventually become slaves to God (Romans 6:22). We surrender ourselves more fully and allow Him to set us aside for His purposes.

Righteousness

While we're here on earth, righteousness isn't a quality we possess but a path we've chosen. While I may not be righteous in every action, I pursue a righteous lifestyle.

God's discipline often pushes us into more righteous behavior. Many years ago, a friend I'll call Linda became disenchanted with me. She was going through severe trials in her own life. Through no fault of mine, my presence in her life seem to aggravate these trials. I tried to be sensitive to her needs but I sometimes failed. I often apologized to her. Even when I thought I was doing right by her, she lashed out at me.

I felt pushed into righteousness to keep my nose clean: I didn't gossip about her; I explained things calmly to her; I prayed that the situation would work out for her; I prayed for myself that I wouldn't become annoyed with her; when people criticized her, I defended her. I was determined to come away from the situation as clean as possible. I look back now and am amazed at how hard I tried to obey God. Afterward, it was easier not to gossip, not to be annoyed with people who were upset with me because of my experience with Linda.

Our trials can push us into righteousness even when we aren't willing. After a heart attack or death, people often say, "I know what's important now." These crises force us to fix our eyes on eternal things and our perspective on life becomes clearer.

Peace

As long as we're open to God's grace during times of discipline, we can find a heaven-centered peace. We moved recently and it was an action-packed two months. We needed to use the money from the house we had owned as a down payment for the one we wanted to buy. At first, the couple buying our house and the investor who owned the one we wanted to buy agreed on a date to close dealings. Then the couple changed their minds and the uproar began. Changes of address had to be rescinded and the investor had to be notified. He threatened to sell the house out from under us even if we paid him $50 a day. He told us to force the other couple to comply.

Every day was filled with dramatic telephone calls followed by an impasse in which we knew nothing. In the beginning I felt defeated. I kept hoping that each call from our real estate agent would bring peace. On occasion, I could be muttering, "So, it's no big deal if I don't know where I'll be living in 30 days."

As I sat down for my noon devotions about two weeks into this calamity, I looked across at my bookshelf (the one I hadn't packed yet), and spied the biography of Hudson Taylor, missionary to China. I thought about how he'd relied on faith. He often didn't know from day to day where God would send him within the interior of China since missionaries didn't normally venture that far yet (Marshall Broomhall, *Hudson Taylor The Man Who Believed God,* Edinburgh: R.&R. Clark, 1929, pp. 63-64). I thought about how I nursed a secret desire within myself to abandon all my middle class trappings and follow in Taylor's footsteps. *Ha,* I thought, *I can't even survive not knowing where I'll live in a month. How could I make it if I didn't know what would happen tomorrow?*

It was at this point that I tasted something of rejoicing in my sufferings. *This is my chance,* I thought, *to trust God when I really don't know what's going to happen.* None of the strong-willed people around me could agree. If they didn't cooperate soon, my family, my office, and I would have to float around among friends for a while.

I know it sounds silly because it's nothing like the way that Hudson Taylor lived without "a safety net." But it was my way of walking in Taylor's footsteps, training myself to live in complete abandon to God.

I pledged myself, like a young Brownie scout, to rejoice in not knowing my future. As it turned out, I didn't exactly rejoice, but I didn't panic either. At times, I felt a glimmer of excitement that I could know a bit of what it was like to walk by faith.

When the verdict came in (both parties gave in a little), I was almost sorry. I'd gotten used to believing in God's power to direct my steps. Knowing the future seemed a little boring.

I view this peace I had in the midst of turmoil as a great gift. Many times before, I'd become agitated with God during crises. This time, I caught on more quickly that He was in charge and that I could trust Him. I learned something of "resting" in God's presence. I'm hoping that I'll catch on more quickly next time. That peace protected me against the temptation to panic, to get mad, to feel sorry for myself. I saw how the peace of God could guard my heart and mind (Philippians 4:7).

Trials then teach us to share His holiness because we learn to love Him no matter what. They teach us righteousness, because we learn to obey Him no matter what. They teach us peace, because we learn to trust Him no matter what.

Is This Where Joy Comes In?

To me, some of the most mysterious and illusive commands in Scripture have to do with rejoicing during trials. I have only had a few experiences with this myself, but I trust that it's true because the Bible says: "But rejoice when you participate in the sufferings of Christ, so that you may be overjoyed when His glory is revealed" (1 Peter 4:13); "Consider it pure joy, my brothers, whenever you face trials of many kinds because you know that the testing of your faith develops perseverance" (James 1:2-3).

The joy these verses talk about comes from our identification with Christ, from knowing that His glory will be revealed. The present may be bleak, but the ending is glorious. We also become joyful because the testing of our faith causes us to grow, which is one of God's goals for us. So as we push toward God's goals, we rejoice in the progress.

How do we develop that joy in the midst of suffering?

Anticipate what God will do. Instead of tapping our feet impatiently in panic, we wait with a half-smile knowing that God will work it out. He has a plan that is bigger than what anyone else would dream.

This joy isn't exemplified by the almost masochistic person who grins and says, "I get to suffer for Jesus." This is the person who faces trials and thinks,

—This is a chance to see God do those miraculous things I've seen Him do so often before.

—This is a chance for me to grow, to become a more authentic vessel for Christ.

—This is my chance to fathom the depths of knowing Christ, to know the "fellowship of sharing in His sufferings" (Philippians 3:10).

Enjoy your relationship with God without interruption. Have you ever heard someone say, "I was doing great with the Lord, but then this problem occurred"? At their worst, problems cause bitterness. At the least, they distract us from our regular spiritual exercises of delighting in God, praising Him, and being thankful for His provision.

Normally, these spiritual exercises create joy in our lives. This joy is obviously different from happiness over earthly circumstances. It flows out of our worship times as we enjoy God and His glory. It pops up when we discover how blessed we are, when we see someone accept the Lord. When we determine to rejoice in our sufferings, these exercises aren't interrupted and our joy in the Lord continues.

If rejoicing in your trials sounds unreal to you, you're not alone. The first time I read the following quote, it sounded like a foreign language to me. Time and again, I've gone back to it. Now and then I've experienced it to some measure. I long to be like this surf-rider:

> The surf that distresses the ordinary swimmer produces in the surf-rider the super-joy of going clean through it. Apply that to our own circumstances, these very things—tribulation, distress, persecution—produce in us the super-joy; they are not things to fight. . . . The saint never knows the joy of the Lord in spite of tribulation, but *because* of it—"I am exceedingly joyful in all our tribulation," says Paul (Oswald Chambers, *My Utmost for His Highest,* Westwood, NJ: Barbour and Company, 1983, p. 48).

It's not a goal you hear about often, but it's one to work toward when it hurts to grow.

FOLLOWING DOCTOR'S ORDERS

Make a list of your character strengths, leaving several blank lines between each one. These might be generosity, organization, endurance, or tenderness. Then to the right of each one, write how they developed. Some are probably results of your upbringing and the good qualities your parents had. Others you've learned in the "school of hard knocks," or the "discipline of the Lord." Note these especially and then write a prayer of thanks for your "hard knocks" experiences. Tell someone about at least one of these.

CHARACTER STRENGTHS	HOW THEY DEVELOPED	PRAYER OF THANKS

LEADER'S GUIDE 1

Objective
To help group members gain encouragement from Christians, both historical and contemporary, who have suffered and conquered.

Personal Preparation
□ Read Hebrews 12:1-17, noting the ways that today's objective is reflected in it.
□ Do one of the following: Reread the accounts of biblical persons you admire who suffered and conquered; or chat with a fellow Christian or two who has suffered and conquered. Ask them about the tools they used to persevere. Be prepared to share this as God leads.
□ After you follow the suggestions in *Following the Doctor's Orders,* be prepared to show your drawing to the group to help them feel comfortable about sharing theirs.

Leading the Group
After you've asked a question or shared your own answer, don't be afraid of silence. A good question requires thought.
□ Question 2. Point out that all of the witnesses had faith, looked to the future heavenly realms, and admitted they were only strangers and aliens on earth (11:13).
□ Question 3. Note that we can adopt an "alien and stranger" mentality by not becoming attached to our possessions, popularity, or man's values.
□ Question 6. Use this example to get them started: **We have to tell additional lies to keep from being found out from the first one.**
□ Question 7. Mention the following if other group members don't: television, well meaning friends who gossip, reading romance novels in place of relating to spouses, volunteering too much at church to escape poor relationships at home. The phrase, "throw off," tells us that it will require great force and perseverance.
□ Question 9. Explain that sometimes we don't want to work through the trial or we're mad at God and unwilling to turn to Him for help.
□ Question 10. As group members answer the last part of this question, listen and focus on them as much as their answers. What are the reactions? Does your group understand that Christian growth can hurt? Are there any false notions such as, "If I were in God's will I would be happy." Make some notes about these and address them gently throughout the study. It might be better not to comment at this point.

LEADER'S GUIDE 2

Objective
To help group members focus on the hope of heaven, especially when they're hurting.

Personal Preparation
☐ Read Hebrews 12:1-17, noting the ways that today's objective is reflected in it.
☐ Take a few minutes each day to meditate on Hebrews 12:2. Focus on the thoughts Jesus might have had as He hung on the cross. What could He have been reflecting to God?

Leading the Group
☐ Question 1. We can "fix" our eyes on Jesus by memorizing portions of Scripture, meditating on Scripture, journaling about it, reflecting on Scripture plaques and Christian art.
☐ Question 2. Christ matures our faith as He leads us into maturity. Eventually He will lead us to enter heaven and dwell with Him there.
☐ Question 4. Jesus could look forward to being exalted to the highest place, given a name above every name, and seeing every knee bow and tongue confess that He is Christ.
☐ Question 5. Jesus was probably hurt by: the testimony of false witnesses (26:60-61); being accused of breaking the Law through blasphemy (26:65); the priests spitting on Him and tearing His clothes (26:67); Peter denying Him (26:69-75); Judas hanging himself (27:1-10); the crowds' choice of Barabbas (27:21); being flogged, stripped, and mocked (27:26, 28-31) by soldiers; appearing naked on the cross (27:35 and historical records of crucifixion customs); insulted (27:39); the agony of being forsaken by God (27:46); John 20:25 also tells us that He was nailed to the cross.
☐ Question 6. When asking for descriptions of Christ's throne, ask anyone who drew their impressions to share them.
☐ Question 10. Stephen was full of the Holy Ghost. Ask your class to tell what they think that means.
☐ Question 15. Abel, Samson, "others," and Jesus shed blood and even died (11:4, 32, 35-38; 12:2). Part of the reason Jesus stayed so focused was that He spent time cultivating the inner life: prayer, solitude, fasting.

LEADER'S GUIDE 3

Objective
To explore how the discipline of the Lord is the act of a loving, caring parent.

Personal Preparation
☐ Read Hebrews 12:1-17, noting the ways that today's objective is reflected in it.
☐ Recall comments by the women in your group about their relationships with their parents. If they've had harsh parents, expect this session to be more difficult to understand. Stress how God's discipline is balanced and bathed in love.
☐ As you study Hebrews 12:5-10, think about your own relationship with God and how He has been a Parent/Disciplinarian to you. Pray for understanding so that you can help other group members.

Leading the Group
Before you answer the questions, emphasize that the word *son* in this passage refers to offspring. The words *child* or *daughter* can correctly be substituted for *son.*
☐ Question 2. They had forgotten God's word of encouragement (v. 5). Notice how the writer of Hebrews mentioned twice that they shouldn't lose heart (vv. 3, 5).
☐ Question 3. Christ rebuked them for being warm, for thinking they were rich when they were spiritually wretched.
☐ Question 4. God loved the church at Laodicea. He advised them to repent, to see themselves as they really were. Then they would have the right to sit with Him on the throne.
☐ Question 7. No one is exempt from discipline. Everyone undergoes discipline. When we submit to God's discipline, we live. We share in His holiness.
☐ Question 8. Disciplining Israel was painful for God. He often yearned for Ephraim.
☐ Question 9. The tone of this passage does seem to shift. First, God reflected mercy, then justice, then mercy again. This is the paradox of God and His strength. We can trust Him to show the right amounts of each.

LEADER'S GUIDE 4

Objective

To help group members understand that discipline is education, that God is a kind and careful Teacher.

Personal Preparation

☐ Complete your own study of the questions and the narrative.

☐ Think of times that God has "educated" you with trials. Try to remember the feelings you experienced so that you'll be in tune with group members' struggles.

☐ As you pray for each woman in your group, mention that you know of her struggles.

Leading the Group

☐ Before you begin, explain that when the group studied Hebrews 12:5-10 in Session 3, we emphasized the Christian's role as God's child being disciplined. This chapter emphasizes God as our Teacher.

☐ Question 1. Set a tone of candor by giving examples from your own life of how you may have made light of God's discipline or lost heart.

☐ Question 2. Bible study, prayer, and constant thankfulness are just three practices that nurture growth. Encourage them to list as many as possible and then choose one they need to focus on.

☐ Question 3. Proverbs 3:1-10 talks about obeying and trusting God and not being wise in your own eyes. It tells how profitable wisdom and understanding are.

☐ Question 6. Be ready to share your own experiences.

☐ Questions 9-15. Some of your group members may have Old Testament phobia. Help them by approaching it with enthusiasm. Point out that this passage shows God's patience with Israel. He continually took her back.

☐ Question 12. This delay seemed to have helped the nation of Israel know that God meant business. It gave them more time to overcome their fears, to get used to tabernacle worship, to experience life apart from slavery. They learned to trust God as they saw how He kept their clothing and sandals from wearing out. God turned the delay into discipline (Deuteronomy 8:1-6).

LEADER'S GUIDE 5

Objective
To substitute holiness for happiness as a life goal.

Personal Preparation

☐ Read Hebrews 12:1-17, noting the ways that today's objective is reflected in it.

☐ Ask God to reveal to each group member, including yourself, the depth of their desire for happiness in life.

Leading the Group

☐ Question 1. Suggest a few of these spiritual goals to get the discussion started: running with perseverance, fixing our eyes on Jesus, enduring hardship as discipline.

☐ Question 3. Explore the different translations. Have group members read theirs. Note especially *grievous* (KJV); *most unpleasant* (PH); *sorrowful* (NASB); *isn't enjoyable while it's happening—it hurts!* (TLB)

☐ Question 6. Expect group members to have questions about this. In some cases, God granted special protection to Israel and to the early church members. He still seems to do so, for example with the releasing of speaker Corrie ten Boom from a Nazi concentration camp just before her group was executed. We're not certain how often Christians fall into these special circumstances, but we know that it doesn't always happen. Many missionaries have been slaughtered over the years. Calling these verses to mind when we have troubles is counterproductive because we may not be in one of God's "special circumstances," as Corrie ten Boom seemed to be.

☐ Question 7. Point out that these passages connect discipline and joy directly. Writing to the Philippians from the discipline of a dingy prison cell, Paul's joy seemed to be more of an act of the will than anything. Jesus spoke the words in John 16:22 just before His crucifixion. It carries a more complete perspective to it. You may grieve, but in God's time you will rejoice.

☐ Question 9. Habakkuk knew that God is just and pure—how could He allow an even more wicked and pagan nation to conquer His own people? Don't read too much into the tone of Habakkuk's complaint. It's safe to say that he was firm, but not necessarily rebellious or stubborn.

☐ Question 12. My single friend paraphrased this passage this way: "Though husbands be scarce, jobs not plentiful and apartments expensive, I will trust the Lord to provide everything I need."

LEADER'S GUIDE 6

Objective
To recognize the attitudes necessary to surrender ourselves to God and endure His discipline.

Personal Preparation
☐ Read Hebrews 12:1-17, noting the ways that today's objective is reflected in it.
☐ Pray for group members, especially self-sufficient ones, that they'll be open to surrendering control of their lives.
☐ Read *Binding Up Wounds* at least twice to make sure you've digested it. You may receive more than the usual amount of questions.

Leading the Group
☐ Question 5. When we ask, "Why me, God?" we may be sincerely searching for an answer or we may be suspecting God of being capricious and mean. It depends on how we ask and what our attitude is toward God.
☐ Question 7. God did consider Job blameless and upright, but it's presumptuous and perhaps dangerous for a human to consider himself righteous. It dims the light of the only One who was truly righteous (Romans 3:10-11).
☐ Question 11. God's questions and comments showed Job that humans do not understand God's ways. Even though Job was a righteous man, his comments about God's behavior were foolish.
☐ Question 12. When Job repented, he knew nothing of his future prosperity. When we don't understand God's reasons, many of us find it helpful to try to figure out what God is trying to accomplish. A better path would be to offer a prayer such as, "I don't understand what You're doing, Lord, but I do trust You."
☐ Question 13. Think of ways that these promises of God's unfailing love—that He will help us with temptation, that He will clothe us in garments of salvation (to save us), that He gives us the Holy Spirit, that Christ will return—have helped you and share with the group.
☐ Question 14. Verses 24-25, 27 imply living a life of righteousness. Being lame can be painful and we often use unrighteous ways to manage pain—yelling at others, self-indulgent habits.
☐ Question 15. God may disable us when we flagrantly sin. David had committed adultery with Bathsheba. Judah had disobeyed God by worshiping idols.

LEADER'S GUIDE 7

Objective
To identify the process and results of becoming bitter toward God.

Personal Preparation

☐ Read Hebrews 12:1-17, noting the ways that today's objective is reflected in it.

☐ Pray especially for new Christians or young Christians in your group. Because of their lack of life experiences or rose-colored opinions of God, they may not be able to understand yet how Christians can become bitter toward God.

Leading the Group

☐ Question 1. It's difficult to live in peace with people who ask us to do things that are dishonest or disrespectful. Yet we stay at peace with ourselves because we don't expect everyone to agree, and we care more about what God thinks than what they think.

☐ Question 2. Holiness refers to setting ourselves apart for God and committing ourselves to a path of righteousness, not in being perfect. "Seeing" the Lord may refer to His second coming or simply to intimate fellowship with God on earth.

☐ Question 3. Grace is God's love and mercy toward those who don't deserve it. Experiencing the discipline of the Lord can make us feel that God doesn't love us. Stress that this happens even to Christians.

☐ Question 4. These people may be angry at God because of the trial and choose to reject or ignore Him.

☐ Question 5. God gives grace to the humble. Pride blocks our sense of worship and mystery of God. As in the Garden of Eden, we think we can be as smart as God. We become bitter because we judge His actions and decide He was in error.

☐ Question 10. Explain the word *despised* this way: **This birthright was a treasure of great value. It's comparable to tossing aside a million dollar diamond ring that's been in your family for years. To throw it aside is to "despise" it.**

☐ Question 12. Bitterness distorts our viewpoint. Esau sold Jacob his birthright, but in bitterness he accused Jacob of stealing it.

☐ Question 14. Esau wanted his blessing so much that he didn't consider that it might be God's will that Jacob have it instead. He could have sought God's purpose for himself or he could have allied himself with Jacob so as to receive benefits from the blessing.

LEADER'S GUIDE 8

Objective

Responding correctly to the discipline of the Lord results in righteousness, peace, and sharing in God's holiness.

Personal Preparation

☐ Read Hebrews 12:1-17 and meditate on it. How has it rounded out your view of God's dealing with you?

☐ Review your own experiences with the discipline of the Lord. How have they resulted in righteousness, peace, or sharing in God's holiness in your life?

Leading the Group

☐ Question 1. This passage mentions the spiritual qualities of righteousness, peace, and "sharing in His holiness." This study will investigate these three results of God's discipline.

☐ Question 2. The left side of the chart should include at least these three characteristics: avoid sexual immorality; control body in a holy and honorable way, not in passionate lust; not wrong someone or take advantage of him.

☐ Question 3. Here are some suggested answers.

Philippians 3:10	*phrase:* fellowship of sharing in His sufferings
	result: attain the resurrection of the dead
1 Peter 4:13	*phrase:* participate in the sufferings of Christ
	result: overjoyed when His glory is revealed
2 Peter 1:4	*phrase:* participate in God's eternal nature
	result: escape corruption of the world

☐ Question 8. The qualities mentioned can be found in vv. 1-2 (dependence on God); vv. 3, 10 (loved by God); vv. 5-6, 9, 11 (praise for God); v. 7 (faithfulness to God); vv. 4, 6 (horror at his own predicament).

☐ Question 10. David showed righteousness: he didn't kill Saul and he prevented his men from doing so. He even felt conscience stricken for having cut off the corner of Saul's robe. David also showed peace: he seemed to be at peace that he wasn't king, even though he had been anointed. Saul was God's anointed king and David refused to interfere.

☐ Question 11. Although holiness includes pure and devoted living, the holy person still sins and shouldn't be overly discouraged by it. We can take comfort that we are part of a "chosen people, a royal priesthood, a holy nation" (1 Peter 2:9) even when we lack faith and obedience.